Algorithms and Combinatorics 11

Klaus Voss

Discrete Images, Objects, and Functions in Z^n

With 100 Figures

Springer-Verlag
Berlin Heidelberg New York
London Paris Tokyo
Hong Kong Barcelona
Budapest

Klaus Voss
Friedrich-Schiller-Universität Jena
Mathematische Fakultät
Lehrstuhl Digitale Bildverarbeitung
UHH, 17. OG
O-6900 Jena, Deutschland

Mathematics Subject Classification (1991): 68T10, 68U10, 68R10

ISBN 3-540-55943-4 Springer-Verlag Berlin Heidelberg New York
ISBN 0-387-55943-4 Springer-Verlag New York Berlin Heidelberg

Library of Congress Cataloging-in-Publication Data
Voss, K. (Klaus): Discrete images, objects, and functions in Z^n / Klaus Voss. p. cm. –
(Algorithms and combinatorics ; 11) Includes bibliographical references and index.
ISBN 0-387-55943-4
1. Image processing–Mathematics. 2. Topology. I. Title. II. Series. TA1632.V6714 1993
621.36'7'0151–dc20 92-45827

Printed in the United States of America

Typesetting: Camera-ready by author
41/3140 - 5 4 3 2 1 0 - Printed on acid-free paper

Preface

Man wird dem einzelnen nicht gerecht, wenn man es gesondert ins Auge faßt, ohne seinen Zusammenhang mit dem Ganzen zu beachten und dem Beziehungssystem Rechnung zu tragen, in dem es steht.

Thomas Mann

Science in general, as well as in each of its individual fields, is a part of human culture. In that sense, this book aims to contribute to uncovering a small part of the connections and relationships which bind image processing, categorized in informatics and technology, with the knowledge accumulated over the years on discrete structures.

How does one consider problems, models, mathematical methods and practical applications? How does the search for ideas and the endeavour for knowledge in the original work of scientists find expression? Is there something to be learnt from science to date for future developments? Such questions have shaped the content and style of this book.

Substantial impetus to the discrete theory of image processing was afforded by the work of Rosenfeld and colleagues. Other fruitful sources of ideas considered here are number theoretical problems (Gauß, Minkowski) and integral geometric investigations (Blaschke, Santalo). Since the beginning of the 1980s I have strived to build upon these ideas a unified mathematical representation of discrete image processing working together with R.Klette and P.Hufnagl.

In a series of lectures, in many individual projects and in a monograph I have on the one hand dealt with special problems, and on the other hand I have striven to form a whole from the puzzling parts.

In the end the following scheme developed: at the beginning is a discrete topology (instead of the classical set theoretical topology) in the form of the largely combinatoric theory of the n-dimensional incidence structures. On this foundation the various mathematical disciplines of set theory ($n=0$), graph theory ($n=1$), and the theory of oriented neighborhood structures ($n=2$) can be built upon one another.

Finally, especially fruitful for image processing is the consideration of the n-dimensional number lattice Z^n as a special incidence structure and the investigation of discrete objects, i.e. of subsets of Z^n.

The introduction of an Euclidean metric into the number lattice links number theoretical methods with geometrical methods (digital straight lines, convex objects, digitalization effects etc.). Because image processing and computer graphics come into very close contact in this field, the discreteness of the number lattice can also lead to new algorithms of computational geometry.

If one assigns numerical values to the points of Z^n, one obtains discrete functions whose investigation forms the last part of the book. Instead of the symbols $\int$ and $\varepsilon \to 0$ of classical analysis, we use here the symbols Σ and $\Delta k = 1$. Several examples will show that the discrete viewpoint also allows the development of new algorithms for practical image processing.

Set theory, graph theory, geometry, combinatorics, and algebra provide the necessary mathematical tools. Certainly, these disciplines are only partially handled during university studies. But since there are sufficient textbooks on these subjects, I have endeavoured to be concise in the compilation of the necessary formulas and laws. However, this book should not be seen primarily as a mathematical textbook since the actual purpose of each theory lies in imparting new insights into the subject and providing new methods for practical application.

The book is a textbook as well as a monograph. Therefore the most important results are cited once more at the end in the form of a compendium. Many problems and connections are only briefly mentioned - partly to stimulate the imagination of the reader but also because the questions can still not be answered today.

The ideas presented here have been stimulated in discussions with many colleagues in the past. Above all I would like to mention U.Eckardt (Hamburg), E.Hertel (Jena), P.Hufnagl (Berlin), A.Hübler (Jena), R.Klette (Berlin), H.Süße (Jena), and W.Wilhelmi (Passau). The manuscript of the book was read carefully and critically by H.D.Hecker (Jena) and H.Süße (Jena), and it was linguistically corrected by R.Baker (Bradford). My very special thanks go to all of them.

Klaus Voss
Jena, Juni 1992

Content

4 Geometrical Laws and Properties

5 Discrete Functions

1 Neighborhood Structures

1.1 Finite Graphs

1.1.1 Historical Remarks

Nearly 250 years ago, a small paper was published by Leonhard Euler [Eu36]. This paper can be considered the birth-certificate of graph theory. Euler stated in a letter from March 1736:

> "Es wurde mir einmal eine Aufgabe über eine Insel vorgelegt, die in der Stadt Königsberg gelegen und von einem Fluß umgeben ist, über welchen sieben Brücken führen, und es wurde gefragt, ob jemand die einzelnen Brücken in einem zusammenhängenden Laufe so durchwandern könne, daß jede Brücke nur einmal überquert wird. Dabei wurde mir auch mitgeteilt, daß bisher weder jemand sich für diese Möglichkeit verbirgt noch jemand bewiesen habe, daß es unmöglich sei, dies zu tun." [Euler, Ko86] [T1)]

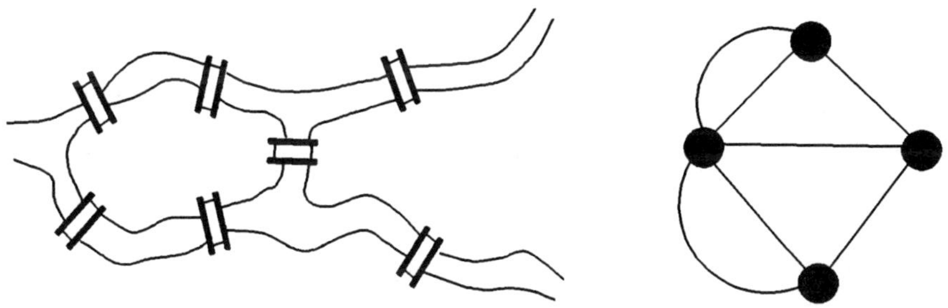

Fig. 1.1.-1: The seven bridges of Königsberg

It was a great achievement to abstract from all the details of river, island, and bridges the simple graph structure on the right of figure 1.1.-1. Today, we can solve a wide range of such problems using the theorem that an open or closed Euler line in a graph only then exists when the graph has exactly two or zero nodes with an odd number of outgoing edges.

Precisely 200 years later, D.König had published his book "Theorie der endlichen und unendlichen Graphen" in Leipzig. The Hungarian author had collected at that time almost all relevant publications creating the first textbook

---

[T1)] Translation of German quotations are given at the end of the reference list.

of graph theory [Ko86]. The bibliography in König's book included only 115 entries. A bibliography from 1963 mentions 1600 papers, and in the following five years more than two thousand graph theoretical books and papers appeared [Sa70]. Today, it is no problem to compile a bibliography of graph theory with 138 textbooks, monographies, and proceedings [Ha89].

From the wide field of graph theory, only a small part appears in this book. This part is characterized by a fundamental notion which König pointed out in the preface of his book:

> "Die Graphentheorie läßt sich von zwei verschiedenen Standpunkten aus auffassen. Erstens bildet sie, als Theorie der eindimensionalen Komplexe, den ersten Teil der allgemeinen Topologie. Zweitens läßt sie sich - wenn man von ihrem kontinuierlich-geometrischen Inhalt abstrahiert - als Zweig der Kombinatorik und der abstrakten Mengenlehre auffassen. In diesem Buch soll dieser zweite Standpunkt zur Geltung kommen, hauptsächlich dadurch, daß wir den Elementen der Graphen - den Punkten und Kanten - überhaupt keinen geometrischen Inhalt zuschreiben: die Punkte (Knotenpunkte) sind beliebige unterscheidbare Elemente, und eine Kante ist nichts anderes als eine Zusammenfassung ihrer beiden Endpunkte." [König, Ko86] [T2)]

Of course, we will use clear geometrical conceptions to formulate or to prove theorems and conclusions. But it must be always clear and understood that an "edge" from the viewpoint of informatics is not an infinite continuous set of points but solely a finite set $\{p,q\}$ composed of the two elements p and q.

Exactly in this sense are graphs defined in all textbooks. Some of these "historical" definitions shall be given here (see also definition 1.1.-1):

> "Ein Graph besteht aus einer Menge X von Elementen erster Art, den sogenannten Knotenpunkten, und einer Menge U von Elementen zweiter Art sowie einer auf U erklärten Funktion (der *Inzidenzfunktion*), die jedem Element $u \in U$ eindeutig ein geordnetes oder ungeordnetes Paar (nicht notwendig verschiedener) Knotenpunkte $x,y \in X$ zuordnet." [Sachs, Sa70] [T3)]

> "Ist M eine beliebige endliche oder unendliche Menge und wird jedem (nicht notwendig geordneten) Paare (A,B), das man aus den Elementen von M bilden kann, eine endliche oder unendliche Mächtigkeit $\boldsymbol{m}_{AB}=\boldsymbol{m}_{BA}$, welche auch Null sein kann, so zugeordnet, daß für jedes A wenigstens ein $\boldsymbol{m}_{AB}$ nicht Null ist, dann gibt es einen Graphen, der als Knotenpunkte die Elemente von M besitzt und in dem irgend zwei Knotenpunkte A und B durch $\boldsymbol{m}_{AB}$ Kanten verbunden sind." [König, Ko86] [T4)]

1.1.2 Elementary Theory of Sets and Relations

The definitions quoted in the foregoing section show that the set theory is the basis of graph theory. Because of that we will compile here some of the most important notions of set theory.

- An (unordered, finite) *set* $M=\{p,q,\ldots\}$ is an enumeration of *elements* so that $\{p,q,r,s\}$ is the same set as $\{q,r,p,s\}$ etc. A set without elements is called an *empty set* and denoted by $\varnothing$.

- A *sequence* $F=(p,q,\ldots)$ is an *ordered set* of elements. The successsive order of elements $p,q,\ldots$ characterizes the sequence F. A *tupel* $[p,q,\ldots]$ is a special finite sequence.

- The fact that p is an element of set M will be designed by $p\in M$. Analogically $p\notin M$ means that p is not an element of M.

- The *union* $M\cup N$ of two sets M and N is a set which contains exactly all elements r with $r\in M$ or $r\in N$.

- The *intersection* $M\cap N$ of two sets M and N is a set which contains exactly all elements r with $r\in M$ and $r\in N$. We call two sets M and N *disjoint* if $M\cap N=\varnothing$.

- A *subset* M of the set N contains only (perhaps all) elements of N. We symbolize that fact by $M\subseteq N$. If some elements of N are not in M, we call M a *proper subset* of N and write $M\subset N$.

- The *complementary set* $\overline{M}=N-M$ in respect to N is a subset of N which contains all elements $p\notin M$.

- The number n of elements of a finite set M will be called as *cardinality* of M. We write $n=card(M)$.

- The *Cartesian product* $M_1\times M_2\times\ldots\times M_k$ of k sets $M_1,M_2,\ldots,M_k$ is a set M with composed elements $m=(m_1,m_2,\ldots,m_k)$ whereby $m_i\in M_i$. Especially, M^n is the set of all sequences $(p_1,\ldots,p_n)$ with $p_i\in M$.

The following formulas are important laws in set theory:

$$(P \cup Q) \cap R = (P \cap R) \cup (Q \cap R) ,$$

$$(P \cap Q) \cup R = (P \cup R) \cap (Q \cup R) ,$$

$$\overline{P \cup Q} = \overline{P} \cap \overline{Q} ,$$

$$\overline{P \cap Q} = \overline{P} \cup \overline{Q} .$$

The set theory is also the basis for the theory of relations. Any subset $R \subseteq P^n$ is called an *n-ary relation* in P. *Binary* relations $R \subseteq P^2$ are important for graph theory. When an element $(p,q) \in R$ is given, we say that p and q are related to one another. A binary relation $R \subseteq P \times P$ is called *reflexive* if $(p,p) \in R$ for all $p \in P$. On the other hand, R is called *irreflexive* if $(p,p) \notin R$ for all $p \in P$. A binary relation R is called *symmetrical* if from $(p,q) \in R$ always follows $(q,p) \in R$ for any $p,q \in R$. But if from $(p,q) \in R$ for $p \neq q$ follows $(q,p) \notin R$ for any $p,q \in R$, the relation R is denoted as *anti-symmetrical*. Finally, we call a binary relation R *transitive* if from $(p,q) \in R$ and $(q,r) \in R$ always follows $(p,r) \in R$ for any $p,q,r \in R$.

A reflexive, symmetrical, and transitive relation $R \subseteq P \times P$ is denoted as *equivalence relation*. In this case, we write $p \sim q$ for $(p,q) \in R$. The importance of equivalence relations follows from the connection with the representation of a set P as a union of pairwise disjoint subsets $Q \subseteq P$, i.e. with the *decomposition* of P.

The *equivalence class* C_p of the equivalence relation $R \subseteq P \times P$ is defined by

$$C_p = \{q \mid q \in P , (p,q) \in R\} .$$

That means that C_p consists of all elements $q \in P$ which are equivalent to p. Using the notion of relation, many different mathematical disciplines can be viewed from a uniform standpoint. Besides the equivalence relation, a reflexive, anti-symmetrical, and transitive relation in P is used to describe the arrangement of elements.

The properties of these relations are generally known in the special cases of identity relation $p=q$ or order relation $p \leq q$. On the other hand, an irreflexive and symmetrical relation - called *neighborhood relation* - as a further example of a binary relation is known in graph theory only. We will investigate this relation in detail in the following sections.

1.1.3 Elementary Graph Theory

The notion *graph* means in this book always a finite structure when nothing else is said. For completeness and uniformity of our investigations, we will give a further definition:

Def. 1.1.-1: Given a set P of elements and a set $N \subseteq P \times P$ of pairs (p,q) of elements with $p,q \in P$. The tupel $[P,N]$ is called a *graph*. If we have $(p,p) \notin N$ then $[P,N]$ is denoted as a *loopless graph*. If from $(p,q) \in N$ follows $(q,p) \in N$, we call $[P,N]$ an *undirected graph*. The elements of P are the *points* or *nodes* or *vertices* of the graph, and the elements of N are the *directed edges* of $[P,N]$. If $(p,q),(q,p) \in N$, we denote the set $\{p,q\}$ as an *undirected edge* of the graph (in short, we say also that $\{p,q\}$ is an *edge* of $[P,N]$).

A graph can be represented by drawing points and lines at the plane (see figure 1.1.-2). But in graph theory, such drawings have no argumentative force. The lines need not be straight lines or circular lines, and a crossing point between two lines in the planar representation of the graph is meaningless. Such representations can serve only as heuristical tools for logical and algorithmical proofs. Once more: the graph theory is seen here as a more complicated set theory because the two sets P and N are taken into account.

This non-geometrical meaning of graph theory is clearly expressed in definition 1.1.-2 avoiding the notions *point* and *edge* and introducing the notion of k-dimensional elements. Such an approach is motivated by the intention to generalize "higher-dimensional set theories" (see chapter 2.1).

$P=\{p,q,r,x,,y\}$

$N=\{(p,q),(q,p),(r,q),(q,r),(p,r),(r,p),(r,x),(x,x)\}$

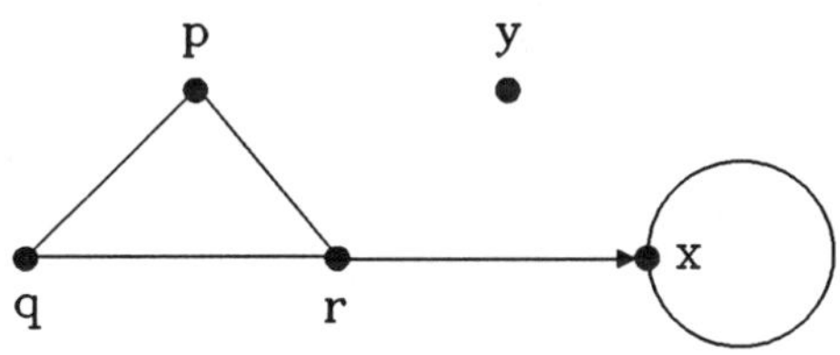

Fig. 1.1.-2: Graphical representation of the graph $[P,N]$

Def. 1.1.-2: Given a set C_0 of so-called zero-dimensional elements and a set C_1 of one-dimensional elements with $C_0 \cap C_1 = \varnothing$. To each $k \in C_1$, exactly two different elements $p,q \in C_0$ are attached by an "attachment" relation $A \subset C_1 \times C_0 \times C_0$. Then the tupel $[C_0,C_1,A]$ is called an undirected, loopless *graph* or a *one-dimensional complex*.

The attachment between elements $k \in C_1$ and $p,q \in C_0$ can be graphically represented, too. In figure 1.1.-3, the elements of C_0 are represented by small circles and the elements of C_1 by small squares. The lines between small circles and small squares represent the attachments between elements of C_0 and elements of C_1. Therefore, we can say that figure 1.1.-3 shows a "hypergraph" (see section 2.1.2).

Both definitions 1.1.-1 and 1.1.-2 are equivalent with $C_0=P$ as set of all points or all zero-dimensional elements and

$$C_1 = \{ \{p,q\} \ / \ p,q \in P \quad , \quad (p,q) \in N \} \ .$$

The number of attachments in figure 1.1.-3 corresponds to the number of lines between elements of C_0 and elements of C_1. This number ist determined by $2 \cdot card(C_1)$. On the other hand, the number of lines is equal to the sum of all line numbers $\nu(p)$ with $p \in C_0$. Now, we will give an exact definition for this fact.

$C_0 = \{p,q,r,x,y\}$

$C_1 = \{k,l,m,n\}$

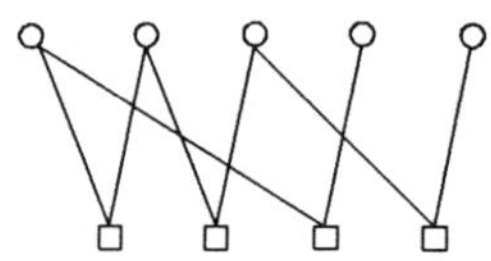

Fig.1.1.-3: Representation of zero- and one-dimensional elements of the complex $[C_0,C_1,A]$

Def. 1.1.-3: We denote by $\nu(p)$ the number of all undirected edges $\{p,q\}$ of a neighborhood graph $[P,N]$ with $p,q \in P,(p,q) \in N$. Likewise, $\nu(p)$ is the number of all one-dimensional elements $k \in C_1$ of a complex $[C_0,C_1,A]$ with $p,q \in C_0,(k,p,q) \in A$. This number is called *node degree* or *neigborhood degree* or *vertex degree* of the element p. An element p with $\nu(p)=0$ is an *isolated element*. A graph or a one-dimensional complex is *regular* if $\nu(p)=\nu=$const for all $p \in C_0$.

Using this definition, we obtain a first fundamental theorem of graph theory which we will call *vertex theorem* or *node theorem*:

Theorem 1.1.-1: In each undirected graph $[P,N]$, the number of edges is given by $\varkappa=card(N)/2$, and $\varepsilon=card(P)$ is the number of vertices. Then

$$\sum_{p \in P} \nu(p) = 2\varkappa \qquad \text{and} \qquad \bar{\nu} = \frac{2\varkappa}{\varepsilon}$$

are fulfilled with $\bar{\nu}$ as mean neighborhood degree.

As in the foregoing section, some notions of theory of undirected graphs shall be compiled which are essential for further investigations (see [Se68, Sa70, Ko86, Ha89]).

- An *edge sequence* $(k_1,k_2,\ldots,k_n)$ is a sequence of directed edges $k_i=(p_i,q_i)\in N$ with $q_i=p_{i+1}$ for $i=1,\ldots,n-1$. The number n is denoted as *length* of the edge sequence.

- A *simple edge sequence* is an edge sequence with $k_i \neq k_j$ for $i \neq j$. In a *closed simple edge sequence*, the periodicity condition $q_n=p_1$ is fulfilled.

- A graph $[P,N]$ for which any two points $p,q\in P$ can be connected with one another by an edge sequence is a *connected graph*.

- A connected graph which contains no closed simple edge sequences is a *tree*. It holds always $\varepsilon-\varkappa=1$ for a tree.

- We assume that in a graph $[P,N]$ all (undirected) edges can be constructed by a closed simple edge sequence. Such a graph is called an *Euler graph*, and the closed simple edge sequence is called an *Euler line*.

- A graph with the maximal number $\varkappa=\varepsilon(\varepsilon-1)/2$ of (undirected) edges is a *complete graph*.

- A *path* $(p_1,p_2,\ldots,p_n)$ is a sequence of (neighbored) points with $(p_i,p_{i+1})\in N$. The number n is the *length* of the path. A *simple path* is a path with $p_i \neq p_j$ for $i \neq j$. A simple path with periodicity $p_{n+1}=p_1$ is called a *closed simple path* or a *circle*.

- We assume that in a graph $[P,N]$ all points $p\in P$ can be passed by a circle. Such a graph is called a *Hamilton graph*, and the edge sequence of the circle is called a *Hamilton line*.

To get an idea of the argumentation in graph theory, we will prove the following theorem:

Theorem 1.1.-2: A finite graph contains an Euler line if the graph is connected, and the neigborhood degrees $\nu(p)$ of all $p\in P$ are even numbers.

To prove this theorem, we see at first that the graph must be connected because a nonconnected graph cannot obtain an Euler line. It follows further that all $\nu(p)$ must be even because for each $p\in P$ an ingoing edge of the Euler line must be followed by an outgoing edge.

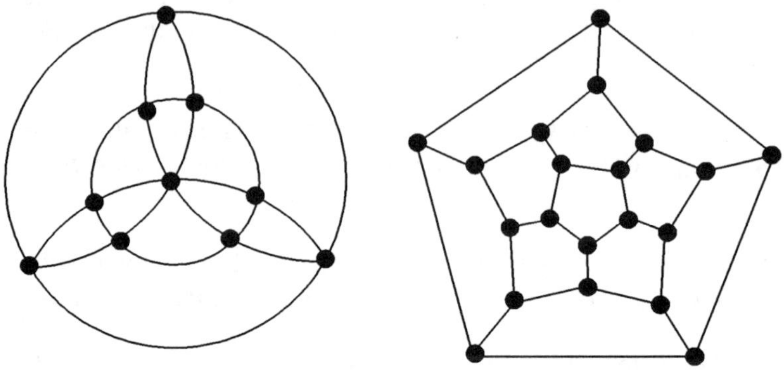

Fig.1.1.-4: Find an Euler line in the left graph and a Hamilton line in the right graph

Vice versa we assume that the graph is connected, and that all $\nu(p)$ are even numbers. Let $S=(k_1,k_2,\dots,k_n)$ be a closed simple edge sequence. Such a sequence always exists because each directed edge $k_i=(p_i,q_i)$ is followed by a directed edge $k_{i+1}=(q_i,r_i)$ - remember that each passage through a point p_i reduces the number of unpassed undirected edges $\{p_i,q\}$ by 2 - and because the graph is finite, the sequence S must be closed.

Either S is already the Euler line of the given graph or not. In the second case, there are further unpassed undirected edges. Be l such an undirected edge $\{p_i,q\}$ at the point p_i. Then a further closed simple edge sequence $S'=(l_1,\dots,l_m)$ with $l_1=l=(p_i,q)$ can be found. Now, $S''=(k_1,\dots,k_n,l_1,\dots,l_m)$ is a longer closed simple edge sequence. Either S'' is the Euler line or not. In the second case...etc.

Contrary to the relatively simple existence conditions of Euler lines, it is not possible to give a simple decision criterion whether a graph has a Hamilton line or not, and there is no effective algorithm to detect such lines in a graph.

1.2 Neighborhood Graphs

1.2.1 Graph Theory and Image Processing

Digital images are special data structures which need an adapted theory for effective algorithmical processing. It is important and meaningful that the individual data of a digital image are not incoherent but form a whole. One has to take into account for computational processing that, for example, the two picture points $p=(x_p,y_p)=(10,5)$ and $q=(x_q,y_q)=(10,6)$ are "neighbored" to one another whereas the picture point $r=(x_r,y_r)=(70,33)$ is far from the first two points.

For investigation of such neighborhood relations, graph theory has been used already in the beginning of image analysis [Ro70, Pa71]. Later, graph theory was employed also as a tool for analysis and processing of digital images [Pa82]. In all of these investigations, the combinatorical character of graphs is used tacitly within the meaning of König's book: the edge attached to two picture points (pixels) is not composed of "other points". The edge is only defined and characterized by both its end points.

Suppose that a digital image is stored in computer memory. The computer can access to pixel informations. But where in the computer is the "connecting line between two pixels" or where is the "small square built up by four pixels"?

Many practical applications of image processing use such notions as "neighbored points", "neighborhood of pixels", "surrounding of a pixel" etc. If we abstract from all unnecessary ingredients of these notions, only one of the following two statements are left: "The points p and q are neighbored" or "The points p and q are not neighbored". Statements like these express connections between elements of sets. They can be described by relations introduced in section 1.1.2.

Two essential properties characterize the popular notion of neighborhood of picture points. First, a pixel is not neighbored to itself, i.e. the pair (p,p) is no element of the wanted binary relation N. On the other hand, the point p shall be always a neighbor of point q if q is neighbor of p, i.e. from $(p,q)\in N$ follows $(q,p)\in N$. Therefore the neighborhood relation $N\subset P\times P$ is irreflexive and symmetrical. By introduction of relation N, a structure is built up in the set P of picture points.

Def. 1.2.-1: A tupel $[P,N]$ with P as a nonempty set of points and $N\subset P\times P$ as an irreflexive, symmetrical relation is a *neighborhood structure*. A neighborhood structure is a loopless, undirected graph without multiple edges (*neighborhood graph*).

Because of this definition, the theory of neighborhood structures can be understood as a branch of graph theory or as a subbranch of the theory of sets with only one binary relation. However, it is factually motivated that we don't develop a theory of loopless, undirected graphs but the theory of neighborhood structures. On the one hand, the neighborhood structures will prove to be special cases of the more general incidence structures (see part 2).

On the other hand, many questions and problems of image analysis are so particular that they are not mentioned in textbooks of graph theory. And most graph theoretical investigations allow hardly immediate applications in image processing.

This is analogous to the connection between mechanics of systems of mass points and the theory of systems of differential equations. We use differential equations in mechanics and graph theory in image processing. But mechanics is not a branch of higher mathematics, and image processing is not a part of graph theory.

The theory of neighborhood structures has developed since 1985 in a series of papers [Vo85, Vo86, Vo86a, Vo87, Kl87, Kl87a] and has been summarized in a textbook [Vo88]. It was realized also in this time that one has to go beyond the graph theory to develop a full theory of digital images and digital objects in higher dimensional spaces.

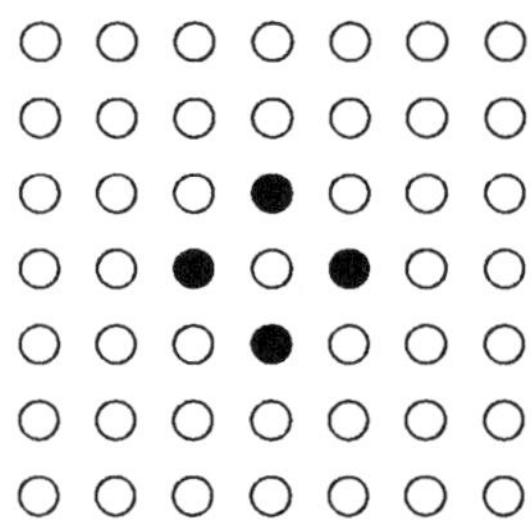

Fig. 1.2.-1: Illustration of neighborhood paradox in binary images

The notion of neighborhood is connected with the terms "4-neighborhood" and "8-neighborhood" for the two-dimensional orthogonal lattice. The first mention of these terms is more than twenty years old [Ro66, Du67]. In binary images, the 4-neighborhood was used and is used to describe the connection of black pixels and the 8-neighborhood to describe white pixel connections, respectively:

> "The reason for this at first sight rather bizarre decision was to avoid anomalies such as those pointed out in [Ro66]: in figure 1.2.-1 if 4-adjacency is used for all pairs of points then the black points are totally disconnected but still separate the set of white points into two components, while 8-adjacency is used for all pairs of points then the black points form a simple closed 8-curve but they do not separate the white points." [Kong and Rosenfeld, Ko86]

> "The 'paradox' can be rephrased as follows: If the 'curve' of (black) points is connected ('gapless'), it does not disconnect its interior from its exterior; if it is totally disconnected, it does disconnect them. This is of course not a mathematical paradox, but it is unsatisfying intuitively." [Rosenfeld and Pfaltz, Ro66]

This argument is very wide-spread in the scientific literature [Ro76, Pa77, Ko84, Ka88]. The paradox is thought to be that a contradiction appears to *Jordan's curve theorem* for the Euclidian plane according to which a closed curve dissects the plane in two disconnected parts.

But in the above quotation, it is not said what we have to understand by a "simple closed 8-curve" in the lattice. Rather it is tacitly assumed that the black points are connected by a Jordan line.

The paradox only appears because in image processing one dispenses with the suppositions necessary for Jordan's theorem but employs this curve theorem for further conclusions. Pavlidis wrote in 1982:

> "When we discuss an image in terms of pixels, we obtain consistent definitions if we use d-connectivity for white pixels and i-connectivity for black pixels. This choice has been popular in the literature but it fails when we deal with pictures containing pixels of more than two colors." [Pavlidis, Pa82]

Kasvand has pointed out in 1988 that there are other difficulties, too:

> "Intrinsically, the regular square raster has two local connectivity patterns during labelling, thinning, erosion, dilatation, etc., depending on whether the foreground or the background of a binary image is being operated upon... When a binary image is negated, and one still wishes to retain the same connectivity in the image, then consistent results are only obtained by keeping track of the sequence of operations applied to the image." [Kasvand, Ka88]

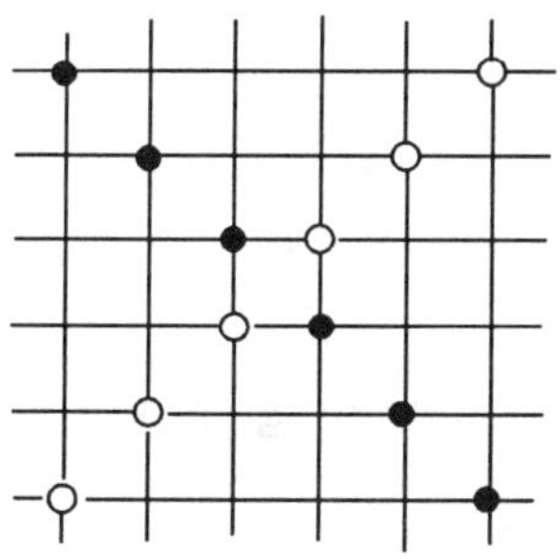

Fig. 1.2.-2: Two digital lines crossing one another

The occurence of paradoxa and difficulties is not astonishing because no consistent theory is offered which is built up on clearly given axioms and definitions. The introduction of two different kinds of connectivity - in three-dimensional orthogonal lattices even of three different kinds [Ko86a] - which shall be held simultaneously in a discrete space is either inconsistent or extremely complicated and unpractical.

There is also another "paradox" in this connection. Two digital straight lines (see section 4.2.2) with different slopes don't have necessarily a common point (figure 1.2.-2). We have accustomed ourselves to this "fact". Now, what is the reason that black and white "lines" may penetrate one another without intersection, but penetrating black and white "curves" must have intersection points "according to Jordan's theorem"?

Contrary to these problems, the theory of neighborhood structures is uniform, simple, and practical. Especially, it is stated by the *separation theorem* 2.2.-7 as a discrete counterpart of Jordan's theorem that a planar, connected neighborhood structure is dissected in *at least* two unconnected parts by the boundary mesh of a region (see section 2.2.3, [Vo88]). We will show in this book that an exact mathematical treatment of discrete spaces will also give an intuitively satisfying theory.

1.2.2 Points, Edges, Paths, and Regions

Using the binary irreflexive and symmetrical neighborhood relation $N \subset P \times P$, it is fixed for each point $p \in P$ to which other points of P the point p is neighbored. Of course, varying neighborhood relations are possible for a given set P.

Def. 1.2.-2: To each $p \in P$ of a neighborhood structure $[P,N]$, a point set $N(p) = \{q \mid (p,q) \in N\}$ is attached which is called *neighborhood* of point p. The points $q \in N(p)$ are the *neighbors* of p. The point p is an *isolated point* if $N(p) = \emptyset$.

All denotations introduced in section 1.1.3 (edge, directed edge, neighborhood degree, edge sequence etc.) are used also for neighborhood structures. Therefore they are to be found in many papers or books dealing with image processing. They belong to the basic vocabulary in image analysis just as elementary set theory and elementary graph theory are the basic tools in this field.

The factual connections between image processing and graph theory require a systematical treatment of connectivity: Only caused by their joints, a set of "stand-alone" picture points is a picture.

Starting from a point $p \in P$ of a neighborhood structure $[P,N]$, we reach any of p's neighbors $q \in N(p)$ via the directed edge (p,q). Now it is possible to reach a further point r which is the "next but one neighbor" of p. This conception of proceeding from one point to another point shall be specified.

Def. 1.2.-3: A sequence $(p_1,\ldots,p_n)$ of points $p_i \in M$ with $M \subseteq P$ is a *path in M* if successive points p_i and p_{i+1} are neighbored in the neighborhood structure $[P,N]$. The sequence (p) wich contains only one point $p \in M$ is also a path in M. The number $n-1$ of directed edges (p_i,p_{i+1}) between the points of the path is the *path length.*

Def. 1.2.-4: Two points $p,q \in M$ of neighborhood structure $[P,N]$ are *connected in respect of M* if there is a path $(p,\ldots,q)$ in $M \subseteq P$.

The two notions of a path in M and of connectivity in M are illustrated in figure 1.2.-3. It is to be seen immediatly that p and q are connected with one another in M or that p and t are not connected in M. Independent of choice of $M \subseteq P$, there is no path from p to t. It is an other case for the points p and s. There is no path between p and s in figure 1.2.-3, because p is an element of M and s is an element of $\overline{M}$. But if we use another subset $M' \subseteq P$, the points p and s can be connected by a path in M'.

There are some important laws for the *connectivity relation* between two points for a set M. Denoting this relation between two points by the symbol $p \sim q$ because this relation is an equivalence relation, we can formulate the following properties:

Reflexivity: The point $p \in M$ is connected to itself because the sequence (p) is also a path. Therefore we obtain $p \sim p$.

Symmetry: Supposing that there is a path $(p,z_1,\ldots,z_n,q)$ in M from p to q then a path exists from q to p because $(q,z_n,\ldots,z_1,p)$ is also a path in M. Therefore from $p \sim q$ follows $q \sim p$.

Transitivity: Supposing that there are paths in M, $(p,x_1,\ldots x_n,q)$ from p to q and $(q,y_1,\ldots y_m,r)$ from q to r. Then a path in M exists from p to r because $(p,x_1,\ldots x_n,q,y_1,\ldots,y_m,r)$ is a path in M, too. Therefore it follows from $p \sim q$ and $q \sim r$ also $p \sim r$.

The connectivity relation in a neighborhood structure $[P,N]$ in respect to $M \subseteq P$ is characterized by a binary relation V_M:

$$V_M \subseteq M \times M \subseteq P \times P \quad .$$

Because of its reflexivity, symmetry, and transitivity, the *connectivity relation* V_M is an equivalence relation (see section 1.1.2). The equivalence classes of V_M are the *maximal connected* subsets of M.

Def. 1.2.-5: A subset $R \subseteq M \subseteq P$ of the neighborhood structure $[P,N]$ is a *connected subset* if for all $p,q \in R$ the relation $p \sim q$ is fulfilled.

Def. 1.2.-6: A connected subset $R \subseteq M \subseteq P$ of the neighborhood structure $[P,N]$ is denoted as *region*.

Def. 1.2.-7: The equivalence classes of $M \subseteq P$ generated by the connectivity relation V_M are the *components* of M. The equivalence classes generated by $V_{\overline{M}}$ are *complementary components of M*.

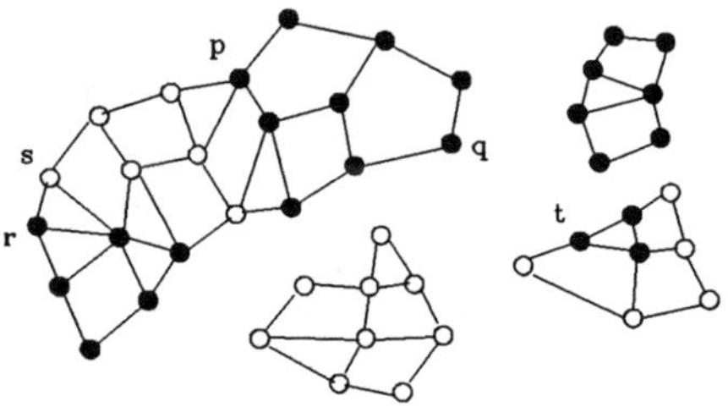

Fig. 1.2.-3: Out of the signed points, only points p and q are connected with respect to set M (●)

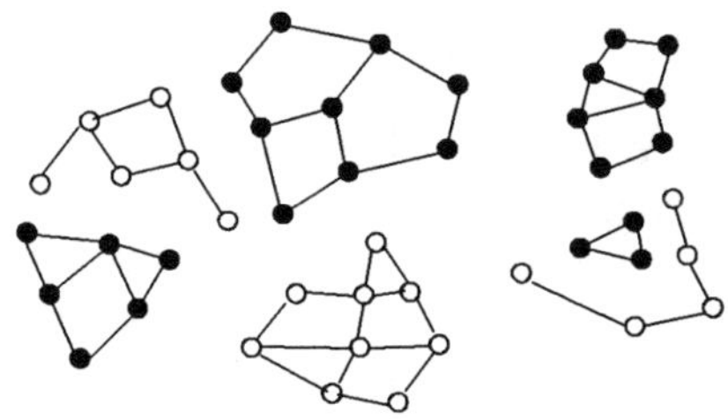

Fig. 1.2.-4: The set P from $[P,N]$ of figure 1.2.-3 is dissected in components and complementary components

In figure 1.2.-4, the components and the complementary components of the set M defined in figure 1.2.-3 are shown. Only the edges (x,y) with $x,y \in M$ or $x,y \notin M$ are keept. It is clear that any two points $p,q \in P$ can be connected in respect of $M \subseteq P$ if and only if they belong either to the same component or to the same complementary component of M.

Now, $C_1,\dots,C_n$ shall be the components of P in respect to the neighborhood relation $N \subset P \times P$. Then we obtain

$$P = \bigcup_{i=1}^{n} C_i \,, \qquad C_i \cap C_j = \emptyset \quad \text{for} \quad i \neq j$$

$$N = \bigcup_{i=1}^{n} N_i \qquad \text{with} \qquad N_i = (C_i \times C_i) \cap N \,.$$

The neighborhood structures $[C_i,N_i]$ are *substructures* of $[P,N]$. There are no connections or relationships between different substructures. Because of that, we can always assume in the following that the neighborhood structure $[P,N]$ to be investigated consists of only one component, i.e. that $[P,N]$ is a connected neighborhood structure.

1.2.3 Matrices of Adjacency

Digital image processing is a "numerical" discipline in the sense that a program running on a computer must be finished in finite time. It is a serious problem to develop effective algorithms - especially time effective algorithms - when we have to treat large data sets as pixel sets of digital images. But in this section, for once we will discuss an ineffective method.

Given a point set $P=\{a,b,\ldots,x\}$ consisting of 24 points. We get a neighborhood structure $[P,N]$ by definition of $N \subset P \times P = \{(a,a),(a,b),\ldots,(x,x)\}$. The concrete neighborhood relation N is given by table 1.2.-1.

If there is a point pair (p_1,p_2) contained in N, the number 1 is put into the *neighborhood matrix* of the table 1.2.-1. Otherwise we put the number 0 into the table symbolizing that by a little point. The irreflexivity of the neighborhood relation N demands that only zeros are on the main diagonal of the matrix, and because of the symmetry of N, the symmetry of the matrix follows [Se68, Ni86, Ha89].

This representation of a neighborhood structure $[P,N]$ as a matrix is unpractical. Who can say whether the structure corresponding to table 1.2.-1 is connected or not?

Def. 1.2.-8: Be given a finite neighborhood structure $[P,N]$ and any order $(p_1,p_2,\ldots,p_n)$ of the $n=card(P)$ elements of P. The $n \cdot n$-matrix A with the coefficients $A_{ij}=1$ if $(p_i,p_j) \in N$ and $A_{ij}=0$ if $(p_i,p_j) \notin N$ is called *neighborhood matrix* or *matrix of adjacency*.

Let A^k be the k-th power of neighborhood matrix A. The coefficients of A are always non-negative. The coefficient A_{ij} has the value 1 or 0 depending whether there is in the neighborhood structure an edge (p_i,p_j) or not. The general coefficient $A_{ij}^{(2)}$ of the Matrix A^2 is obtained by

$$A_{ij}^{(2)} = \sum_k A_{ik} A_{kj} .$$

The terms $A_{ik}A_{kj}$ are 1 or 0 depending whether there is a path (p_i,p_k,p_j) or not. Therefore, the coefficient $A_{ij}^{(2)}$ is the number of all paths from p_i to p_j. Correspondingly, $A_{ij}^{(k)}$ as coefficient of matrix A^k determines the number of all different paths of length k between the point p_i and the point p_j.

If we are interested only in the possibility of whether there is a path from p_i to p_j then the matrix $V^{(k)}$ can be defined by

$$V_{ij}^{(k)} = \min\left(1, A_{ij}^{(k)}\right) .$$

Table 1.2.-1: Matrix representation of a neighborhood structure

	a	b	c	d	e	f	g	h	i	j	k	l	m	n	o	p	q	r	s	t	u	v	w	x
a	.	1	.	.	.	1	.	.	.	1	.	.	.	.	.	.	.	.	.	.	.	.	.	.
b	1	.	.	.	.	1	.	.	.	.	.	.	.	.	.	.	.	1	.	.	.	.	.	.
c	.	.	.	.	.	1	.	.	1	.	.	.	.	.	.	.	.	1	.	.	.	.	.	.
d	.	.	.	.	1	.	1	.	.	.	.	.	.	.	.	.	.	.	.	.	1	.	.	.
e	.	.	.	1	.	.	1	1	.	.	1	.	.	1	.	.	.	.	.	.	1	.	.	.
f	1	1	1	.	.	.	.	.	1	.	.	.	.	.	.	.	.	.	.	.	.	1	.	.
g	.	.	.	1	1	.	.	1	.	.	.	.	.	.	.	1	.	.	.	.	.	.	.	.
h	.	.	.	.	1	.	1	.	.	.	1	.	.	.	.	.	.	.	1	.	.	.	.	.
i	.	.	1	.	.	1	.	.	.	.	.	.	.	.	.	.	1	.	.	1	.	.	.	.
j	1	.	.	.	.	.	.	.	.	.	.	1	.	.	.	.	.	.	.	.	.	.	.	1
k	.	.	.	.	1	.	.	1	.	.	.	.	.	1	1	.	.	.	.	.	.	.	.	.
l	.	.	.	.	.	.	.	.	.	1	.	.	1	.	.	.	.	.	.	.	.	1	1	1
m	.	.	.	.	.	.	.	.	.	.	.	1	.	.	.	.	.	.	.	1	.	1	1	.
n	.	.	.	.	1	.	.	.	.	.	1	.	.	.	1	.	.	.	.	.	1	.	.	.
o	.	.	.	.	.	.	.	.	.	.	1	.	.	1	.	.	.	.	1	.	.	.	.	.
p	.	.	.	.	.	.	1	.	.	.	.	.	.	.	.	.	.	.	1	.	.	.	.	.
q	.	.	.	.	.	.	.	.	1	.	.	.	.	.	.	.	.	1	.	1	.	.	.	.
r	.	1	1	.	.	.	.	.	.	.	.	.	.	.	.	.	1	.	.	.	.	.	.	.
s	.	.	.	.	.	.	.	1	.	.	.	.	.	.	1	1	.	.	.	.	.	.	.	.
t	.	.	.	.	.	.	.	.	1	.	.	.	1	.	.	.	1	.	.	.	.	.	1	.
u	.	.	.	1	1	.	.	.	.	.	.	.	.	1	.	.	.	.	.	.	.	.	.	.
v	.	.	.	.	.	1	.	.	.	.	.	1	1	.	.	.	.	.	.	.	.	.	.	.
w	.	.	.	.	.	.	.	.	.	.	.	1	1	.	.	.	.	.	.	1	.	.	.	1
x	.	.	.	.	.	.	.	.	.	1	.	1	.	.	.	.	.	.	.	.	.	.	1	.

Of course, it is $V^{(1)}=A$. The matrix V with

$$V_{ij} = \min_{1 \le k \le n} \left(V_{ij}^{(k)} \right)$$

describes the connectivity relation in algebraical terms. The coefficients V_{ij} are 1 if there is any path from p_i to p_j in the neighborhood structure $[P,N]$ with $n=card(P)$. The coefficient V_{ij} is equal to zero if there is no such a path.

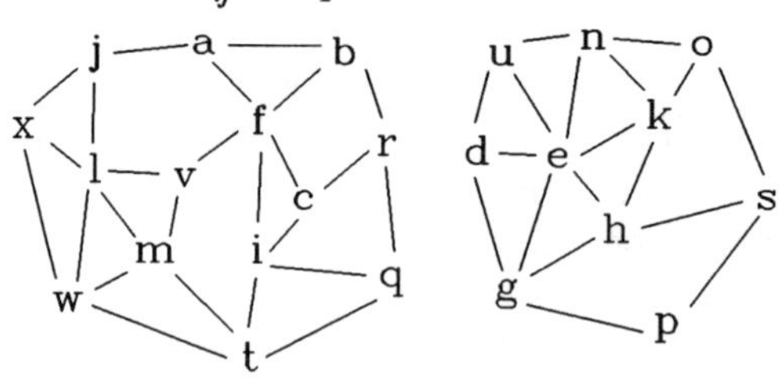

Fig. 1.2.-5: Neighborhood graph corresponding to the matrix of adjacency given in table 1.2.-1

We will point out that the matrix of adjacency is of interest for theoretical investigations. However in practice, we have to develop other methods to determine whether the neighborhood structure given in table 1.2.-1 is connected or not (see figure 1.2.-5).

1.2.4 Graph Distances

The connection between two points in a neighborhood structure is naturally combined with the notion of "distance" between points. Two neighbored points have a small distance from one another while the points *r* and *w* in figure 1.2.-5 are far from one another.

Def. 1.2.-9: A *metric* is introduced in a point set *P* by a real valued *distance* $d(p,q)$ for any two points $p,q \in P$ with the following properties:

$$d(p,p) = 0$$

$$d(p,q) = d(q,p)$$

$$d(p,q) > 0 \quad \text{for} \quad p \neq q$$

$$d(p,q) + d(q,r) \geq d(p,r) \ .$$

The last requirement in this definition is called *triangle inequality*. The Euclidian distance between points in the plane fulfills these conditions (see also section 1.4.1).

Let $l(p,q)$ be the *shortest path length* between the points *p* and *q* in a given neighborhood structure $[P,N]$. Then we can introduce a metric in $[P,N]$ using the distance $d(p,q)=l(p,q)$ which fulfills all requirements of definition 1.2.-9 [Ha89]. Simple examples show already that there may be several different paths with the same minimal path length between points. In figure 1.2.-5, there are two shortest paths (c,f,a,j,x) and (c,i,t,w,x) between *c* and *x*. This fact is a peculiarity in comparison with the Euclidian geometry for which the shortest way between two points *a* and *b* is always the unambiguously defined straight line from *a* to *b*.

Therefore geometries on graphs or geometries on neighborhood structures are characterized by non-Euclidian properties.

It is a typical task of applied graph theory to determine shortest paths [Mi78, Bo81]. First algorithms for that task were developed by Dijkstra [Di59] and Dantzig [Da67].

In a complete graph it is $d(p,q)=1$ for all $p,q \in P$ with $p \neq q$. For each finite connected neighborhood structure $[P,N]$ with $\varepsilon = card(P)$, we obtain $d(p,q) < \varepsilon$. Therefore it is obvious to introduce the following notions [Ha89].

Def. 1.2.-10: We denote the number

$$d_{max}([P,N]) = \max_{p,q \in P} \left(d(p,q) \right)$$

as *diameter* of the neighborhood structure $[P,N]$ or of the corresponding neighborhood graph.

Def. 1.2.-11: For a neighborhood structure $[P,N]$, the number

$$e(p) = \max_{q \in P} \left(d(p,q) \right)$$

is called the *excentricity* of the point $p \in P$. The number

$$r\left([P,N]\right) = \min_{p \in P} \left(e(p) \right)$$

is the *radius* of the neighborhood structure $[P,N]$, and the set

$$C\left([P,N]\right) = \left\{ q \mid e(q) = r([P,N]) \right\}$$

is the *centre* of $[P,N]$.

Def. 1.2.-12: Let M and M' be two subsets of P in respect to the neighborhood structure $[P,N]$. Then the number

$$d(M,M') = \min_{m \in M,\, m' \in M'} \left(d(m,m') \right)$$

is the *set distance* between M and M'.

From $M \cap M' \neq \emptyset$, it follows $d(M,M')=0$ because there is at least one point m with $m \in M$, $m \in M'$ and $d(m,m)=0$. For disjoint sets M and M', the set distance $d(M,M')$ is always positive. But we must point out that the set distance does not determine any metric in the set of all subsets of P in a neighborhood structure $[P,N]$.

1.3 Components in Neighborhood Structures

1.3.1 Search in Graphs and Labyrinths

Many practical tasks of image processing are on principle problems of *graph search*. Pavlidis has remarked about this connection:

> "People are very good at focusing on parts of a two-dimensional display, but computers must proceed in a blind fashion. Therefore, picture traversal algorithms are very important. For discrete images, this corresponds to the traversal of a discrete grid that can be interpreted as a graph whose nodes are pixels, and whose branches connect nodes corresponding to neighboring pixels...In fact, because all data structures used for images are graphs, it is a good idea to start with a review of graph traversal algorithms." [Pavlidis, Pa82]

Now let us consider the image in figure 1.3.-1. Using the threshold $t=6$ for the gray values to be taken into account, the subset of all pixels (x,y) with gray value $g(x,y) \geq t$ can be determined as subset $M \subseteq P$ of the set P of all pixels. Contrary to the simple row-column-structure of figure 1.3.-1, a rather complicated neighborhood graph is obtained (figure 1.3.-2).

4	1	7	8	4	3	2	5	5	2	4	4	2
3	8	6	6	0	7	8	9	5	3	8	4	2
5	4	8	7	3	2	7	8	7	8	8	3	1
2	4	9	0	3	5	5	5	7	4	5	3	0
3	5	5	1	0	8	7	9	8	9	4	1	5
3	5	1	2	8	9	7	7	7	8	9	2	3
4	2	3	7	7	7	5	4	7	8	7	5	0
1	3	1	8	9	5	2	4	2	8	8	4	2
0	4	4	9	8	7	7	8	9	9	7	3	0
2	3	3	8	7	8	7	8	8	7	0	2	5
2	2	5	4	7	5	3	7	7	5	2	3	1
2	3	1	4	0	4	1	0	5	3	0	1	1

Fig. 1.3.-1: Gray value image

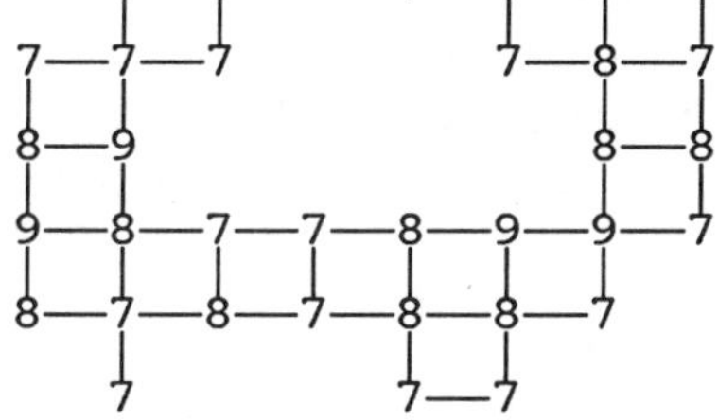

Fig. 1.3.-2: Graph of the central component of figure 1.3.-1

Already the simple question what is the smallest gray value of the central component cannot be answered by a linewise search in the gray value picture. There are further of such typical tasks, for instance there are tasks to determine the

- number of components,
- number of pixels of a component,
- minimal, mean or maximal gray value of a component,
- diameter of a component,
- "structure" of a component.

The question to the "structure" of a component is not precise because "structure" within the meaning of "shape" is not exactly defined. It is to be seen in figure 1.3.-2 that the component has a rather complicated "structure" because there are "holes" and "branchings". It is the aim of the further investigations to give definitions, theorems, algorithms, and methods to solve such questions as asked above.

The problem of graph search is similar to the problem of labyrinth search. A *labyrinth* can be regarded as undirected graph (maybe with loops and multiple edges). We must now find a method to reach outgoing from any point of the labyrinth any other point and any edge - and finally to find again the start point. Yet more difficult, the scheme of the labyrinth is not known.

The first method devoted to the labyrinth problem dates from the classical antiquity. Using the thread of Ariadne, Theseus searched the labyrinth of Kreta, killed the Minotaurus, and achieved exit from the labyrinth.

The graph search requires that all nodes of the graph have to be find out. It would be therefore very effective if one could construct a Hamilton line to pass through each node exactly one by one. But not each graph has a Hamilton line, and there is no effective method to find a Hamilton line in a given graph.

To solve the labyrinth problem, it would be effective to run through an Euler line. The theorem 1.1.-2 determines the conditions for the existence of an Euler line. But not all graphs have an Euler line.

In both cases - graph search and labyrinth search - we need other methods to solve the search problems. Because in image processing the points are decisive, we will investigate only the problem of graph search.

1.3.2 Neighborhood Search

The basic information about the neighborhood graph $[P,N]$ is obtained in the neighborhood relation N. But we have seen in section 1.2.3 that the matrices of adjacency are insufficient tools for practical investigation of graphs.

Lists are distinctly better suited for graph search [Ni86]. The graph of figure 1.2.-5 could be characterized by the list

$$L_1 = (\ (a,(b,f,j)),\ (b,(a,f,r)),\ (c,(f,r,i)),\ldots,\ (w,(x,l,m,t)),\ (x,(j,l,w))\ )\ .$$

The list L_1 consists of sublists which themselves consist of an element $p \epsilon P$ and a list of the neighbors of p.

This representation of a neighborhood structure would be the right method for a list processing language like LISP. However, in numerically determined programming languages, the parantheses must be symbolized additionally or the variable list lengths must be given explicitly.

Supposing that the points of figure 1.2.-5 are given by their x- and y-coordinates. Then the following list can be built up:

$$\begin{array}{lllllll} L_2 & = & 24 & x_a\, y_a & 3 & x_b\, y_b\ \ x_f\, y_f\ \ x_j\, y_j \\ & & & x_b\, y_b & 3 & x_a\, y_a\ \ x_f\, y_f\ \ x_r\, y_r \\ & & & x_c\, y_c & 3 & x_f\, y_f\ \ x_r\, y_r\ \ x_i\, y_i \\ & & & \ldots\ldots\ldots\ldots & & \\ & & & x_w\, y_w & 4 & x_x\, y_x\ \ x_l\, y_l\ \ x_m\, y_m\ \ x_t\, y_t \\ & & & x_x\, y_x & 3 & x_j\, y_j\ \ x_l\, y_l\ \ x_w\, y_w \end{array}$$

Because for each point $p \epsilon P$ we must put two coordinate values into the list, the number $\nu = \nu(p)$ of the neighbors of p, and the $2\nu(p)$ coordinate values of all neighbors $q \epsilon N(p)$, we obtain for the whole list length l by means of the node theorem 1.1.-1

$$l = 1 + (2+1)\varepsilon + 2 \sum_{p \epsilon P} \nu(p) = 1 + 3\varepsilon + 4\varkappa \quad .$$

If we attached in addition a gray value to each point $p \epsilon P$, the list length is enlarged to $1+4(\varepsilon+\varkappa)$. In an $n{\cdot}n$-image, we have $\varepsilon = n^2$ and $\varkappa = 2n(n-1)$. Therefore we need a memory size of $1+4n(3n-2)$ bytes or words.

The determination of single components of a neighborhood structure using this list method is as similarly unpractical as the method of neighborhood matrices. But the following algorithm yields, at least in principle, a solution:

<u>Alg. 1.3.-1:</u> Determination of components in a neighborhood structure by means of lists

1. Be given the list L of the neighborhood graph (either $L=L_1$ or $L=L_2$). The single components of the graph have to be determined.
2. If L is empty then go to step 11.
   3. Now choose any sublist $L_p=(p,N(p))$ out of L and delete this sublist in the main list L.
   4. Generate the component list $C=(L_p)$ and the mark list $M=(L_p)$.
   5. If M is empty then go to step 9.

6. Choose a sublist $L_q=(q,N(q))$ out of M and delete this sublist in the mark list M.
7. Prove for all $r\epsilon N(q)$ whether the sublist $L_r=(r,N(r))$ is still contained in the main list L or not. If L_r is in L then delete L_r in L and put L_r into the both lists C and M.
8. Go to step 5.

9. The component list C describes one component of the original neighborhood graph $[P,N]$. Now, this component can be investigated in detail.
10. Go to step 2.

11. All components of the structure $[P,N]$ are determined and investigated.

Step 7 describes the decisive test whether the neighbors $r\epsilon N(q)$ of a point q are already recorded or not. This test and inserting or deleting of sublists in lists has to be very effective, but cannot be performed effectively in the method described above. Therefore, an other approach to detect components is used in image processing.

In an $n{\cdot}n$-image, the *gray values* are stored at places which can adressed by integer coordinate values x and y with $1\leq x\leq n$ and $1\leq y\leq n$. To decide whether a pixel of a component is already recorded or not, we introduce a *marking* of picture points (x,y) by the function $m(x,y)$. The picture of all mark values can be a simple binary image. The mark values - called "marks" for short - are class signs for picture points.

In $n{\cdot}n$-images, it is not necessary to store the neighbors of a pixel in lists. It is fixed in image processing, for example, that the neighbors of the point (x,y) are

the left neighbor	$(x-1,y)$,
the upper neighbor	$(x,y-1)$,
the right neighbor	$(x+1,y)$,
the lower neighbor	$(x,y+1)$.

Because of that, the neighbors can be determined by their adresses. The special data structure "image" allows an effective access to gray values and mark values [Ni86].

1.3.3 Graph Search in Images

In general, it is necessary to work by means of list also in $n{\cdot}n$-images. At least, we need the list M of pixels still waiting for handling. One possibility to organize this list is the stack technique. A *stack* is a list where a new element can be added only at the end of the list, and only the element added last can be taken from the list (figure 1.3.-3). An access to elements inside of the list is forbidden. Therefore, we call such lists *LIFO-lists* (last-in-first-out).

The adding of a new element to a LIFO-list shall be done by the procedure **PUSH**. Picking out the last added element is realized by the procedure **POP**. The initial (empty) list is organized by the procedure **INIT_STACK**, and the test whether there are further elements in the list or not can be done by the Boolean function **EMPTY_STACK**.

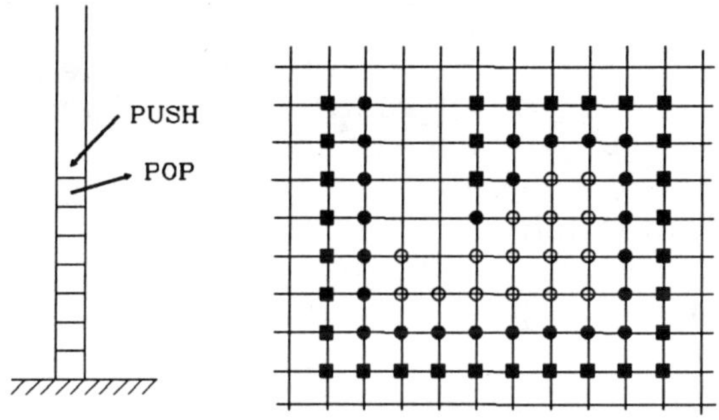

Fig. 1.3.-3: Graph search by LIFO-method (■ points recorded and marked, ● in stack contained, ○ not yet processed)

In the following algorithm 1.3.-2, the number CNUMBER of components and PNUMBER as number of *pixels* (i.e. picture points or picture elements) per component have to be determined. The pixels to be investigated are characterized by the simple decisions $g(x,y) \geq$ THRESHOLD or $g(x,y) <$ THRESHOLD where the function $g(x,y)$ means the gray value of the point (x,y). A pixel (x,y) is then a point of the components if $g(x,y) \geq$ THRESHOLD is fulfilled, the point is not marked, and the point lies within the image. We will call this multiple decision by the simple Boolean function **TEST_POINT**. Adding a point (x,y) to the list (that means storing its coordinate values by means of the procedure **PUSH**), incrementing PNUMBER, and marking the point will be done by the procedure **MARK**.

Alg. 1.3.-2: Determination of components in a gray value image by stack technique (LIFO)

```
0 → CNUMBER
for YS=1 to N do
begin
    for XS=1 to N do
    begin
        if G(XS,YS)≥THRESHOLD and M(XS,YS)=0 then
```

```
            begin
                inc CNUMBER
                INIT_STACK
                PUSH XS YS
                1 → PNUMBER
                1 → M(XS,YS)
                while not EMPTY_STACK do
                begin
                    POP X Y
                    if TEST_POINT(X-1,Y) then MARK(X-1,Y)
                    if TEST_POINT(X,Y-1) then MARK(X,Y-1)
                    if TEST_POINT(X+1,Y) then MARK(X+1,Y)
                    if TEST_POINT(X,Y+1) then MARK(X,Y+1)
                end "of list processing"
            end "of a component"
            print(CNUMBER,PNUMBER)
        end "of search in a line Y=const"
    end "of search for start points in the image"
```

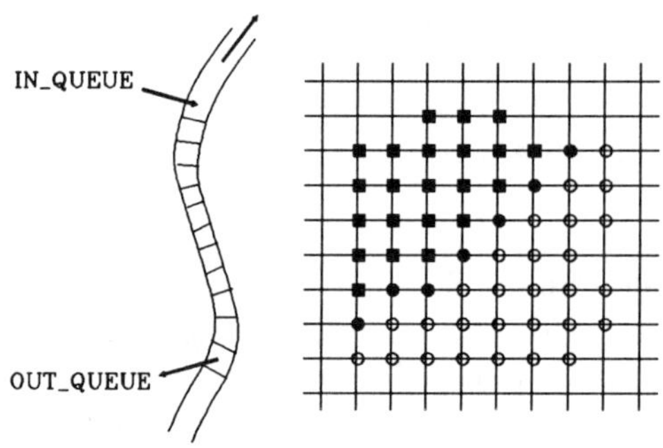

Fig. 1.3.-4: Graph search by FIFO-method (■ points recorded and marked, ● in stack contained, ○ not yet processed)

The disadvantage of the simple-to-implement LIFO-method is that the stack list can be very extended. In figure 1.3.-3 for example, a nearly square-like component is to be investigated, and the sequence of the first processed pixels is given with respect to the left-upper-right-lower order of neighbors to be proved. We see that the number of until now not processed but pushed points can be large. The stack can contain up to $\varepsilon/2$ points at times when ε is the point number of the respective component. Because of the order of processed points we denote the LIFO-method also as *depth-first-search*.

In many cases, a more memory-effective method is given by a waiting list organization. We speak of a *queue* of elements. In a queue, a new element is added to the list as in the stack technique. But as the next element to be processed that element is used which is the oldest in the queue (FIFO-method, i.e. first-in-first-out, see figure 1.3.-4). Using the additional procedures **INIT_QUEUE, IN_QUEUE**, **OUT_QUEUE**, and the additional Boolean function **EMPTY_QUEUE**, the main part of algorithm 1.3.-2 can be reformulated:

Alg. 1.3.-3: Determination of components in a gray value image by means of FIFO-method (queue method)

```
...
begin
    inc CNUMBER
    INIT_QUEUE
    IN_QUEUE XS YS
    1 → PNUMBER
    1 → M(XS,YS)
    while not EMPTY_QUEUE do
    begin
         OUT_QUEUE X Y
         if TEST_POINT(X-1,Y) then MARK(X-1,Y)
         if TEST_POINT(X,Y-1) then MARK(X,Y-1)
         if TEST_POINT(X+1,Y) then MARK(X+1,Y)
         if TEST_POINT(X,Y+1) then MARK(X,Y+1)
    end "of list processing"
end "of a component"
print(CNUMBER,PNUMBER)
...
```

It is shown in figure 1.3.-4 that in a likewise nearly square-like component the already processed pixels perform a rather compact cluster. The pixels included in the queue are located at the boundary of the cluster. This set of queue-pixels spreads out like a grass-fire. Contrary to the LIFO-method, about $\varepsilon^{1/2}$ elements are maximally contained in the queue. Therefore the FIFO-method is denoted also as *breadth-first-search*.

Finally, it shall be referred to section 2.2.4 where the search in two-dimensional neighborhood structures is investigated, i.e. the search in structures with zero-, one-, and two-dimensional elements.

1.3.4 Neighbored Sets and Separated Sets

The neighborhood of points and the neighborhood method of graph search lead almost inevitably to the notion of neighborhood of sets or components.

Def. 1.3.-1: Two non-empty disjoint sets $M,M' \subset P$ of a neighborhood structure $[P,N]$ are *neighbored* if there are points $p \in M$ and $q \in M'$ with $(p,q) \in N$. The sets M and M' are then *set neighbors* to one another. The set $\mathbf{P}=\{M_1,M_2,\ldots\}$

of sets $M_i \subset P$ neighbored to $M \subset P$ are the *set neighborhood* $\boldsymbol{N}(M)$ of the set M. Two non-empty disjoint sets $M,M' \subset P$ are *separated* if they are not neighbored.

With respect to this definition, two neighbored sets have to be always disjoint. This requirement is a generalization of definition 1.2.-1/2 after which two points p and q can be only then neighbors to one another if $p \neq q$ or $\{p\} \cap \{q\} = \varnothing$. The set neighborhood relation $\boldsymbol{N} \subset \boldsymbol{P} \times \boldsymbol{P}$ is like the point neighborhood relation irreflexive and symmetrical.

Def. 1.3.-2: Let $[P,N]$ be a neighborhood structure and $\boldsymbol{P}$ a set of pairwise disjoint subsets of P. Then $[\boldsymbol{P},\boldsymbol{N}]$ with $\boldsymbol{N}$ as set neighborhood relation is a *set neighborhood structure*.

All up to now investigated properties and laws of neighborhood structures (neighborhood graphs) are fulfilled also for set neighborhood structures [Kl86]. Figure 1.3.-5 shows the segmentation of an image in regions and the neighborhood graph corresponding to this decomposition. This graph is denoted as *region adjacency graph*.

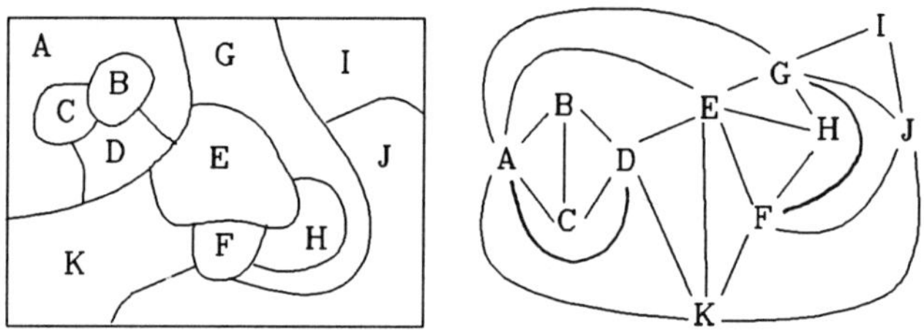

Fig. 1.3.-5: Decomposition of an image in regions and set neighborhood structure of this decomposition

With respect to section 1.2.4, we can say that two neighbored point sets $M,M' \subset P$ of a neighborhood structure $[P,N]$ have the set distance $d(M,M')$. If the distance $d(M,M')$ is larger than 1, then M and M' are separated from one another.

Some theorems are valid only for neighbored and separated sets, not for points (see the relatively simple proofs for these theorems in [Vo88]):

- Two different components of a set $M \subseteq P$ of the neighborhood structure $[P,N]$ are always separated from one another.

- Each component of a proper subset $M \subset P$ of the connected neighborhood structure $[P,N]$ is neighbored to at least one complementary component of M.

- A (connected) region $R \subset P$ of the connected neighborhood structure $[P,N]$ is set neighbored to all of its complementary components.

Theorem 1.3.-1: The complementary set $\overline{C_i}$ of a single complementary component C_i of a point set $M \subset P$ of neighborhood structure $[P,N]$ is always connected.

This *complement theorem* is a necessary condition for investigations in section 3.3.1 where topological equivalence will be discussed. The proof of the complement theorem uses the facts that a complementary component C_i is always a connected point set (i.e. a *region*), and that a region R is set neighbored to all of its complementary components. Therefore, the set

$$\overline{C_i} = R \cup C_1 \cup C_2 \ldots \cup C_{i-1} \cup C_{i+1} \cup \ldots \cup C_I$$

as only complementary component of C_i is likewise always connected.

1.3.5 Component Labeling

In a neighborhood structure $[P,N]$, the set P shall be decomposed in disjoint subsets $M_1,\ldots,M_n$. Now, sub-neighborhood structures are generated by the subsets M_i (see section 1.2.2). We assume that the sub-neighborhood structures are connected. The notion *component labeling* means an operation whereby each point $p \in P$ gets a *mark* or a *label* in a manner that all points of a single component of the neighborhood structure have the same label, but points $p \in M_i$, $q \in M_j$ of neighbored sets M_i and M_j are differently labelled. The component neighborhood graph whose points symbolize components M_i has to be labelled or colored so that neighbored points are differently labelled [Sa72, Vo88, Ha89].

Def. 1.3.-3: A mapping φ from a set P of a neighborhood structure $[P,N]$ into a set C of colors with $\varphi(p) \neq \varphi(q)$ for $(p,q) \in N$ is called *regular coloring* of the corresponding neighborhood graph.

There is always a regular coloring for finite neighborhood graphs because such a set C can be used that each point $p_i \in P$ is colorable by its own color i. But it is of interest to ask how many different colors are sufficient in minimum to color regularly a neighborhood graph.

Def. 1.3.-4: The smallest number $c=card(C)$ of colors by means of which a neighborhood graph $[P,N]$ can be colored regularly is the *chromatic number* of $[P,N]$.

There are a few general valid theorems corresponding to the chromatic number for any graphs which can be easily proved:

- If H is a subgraph of G, then the inequality $c(H) \leq c(G)$ is fulfilled.

- It is always $c(G)=2$ if G is a tree.

But the proofs of the following two theorems are not so simple:

Theorem 1.3.-2: Let H be a graph consisting of only one circle (see section 1.1.3). Then $c(H)=2$ for even number of points is fulfilled, and $c(H)=3$ colors are sufficient for odd point number. A complete graph C_n with n points needs $c(C_n)=n$ colors for regular coloring. Otherwise it is for each neighborhood graph

$$c(G) \quad \leq \quad \max_{p \in P} \left(\nu(p) \right) \quad .$$

Theorem 1.3.-3: It is $c(G) \leq 5$ for each planar graph G.

The theorem 1.3.-2 was proved first by Brooks in 1941 [Br41, Ha89], and the theorem 1.3.-3 was formulated by Heawood in 1890 [He90, Ha89, Vo88], see definition 2.2.-5. Famous and widely known is the four-color-theorem whose history can be retraced up to 1852 [Ri59]. The *four-color-theorem* specifies the theorem 1.3.-3 in the sense that for planar graphs already four colors are sufficient for regular coloring (see section 2.2.5).

Not until 1976 could Appel and Haken prove that the four-color-theorem was right. The pecularity of their proof is that the problem has been solved by means of a computer which needs about 1000 hours of CPU-time [Ap76].

While the four-color-theorem needs planarity of the graphs to be investigated, the theorem 1.3.-2 is fulfilled for any neighborhood graph. Using our knowledge about graphs and neighborhood structures, we can prove that theorem.

- Given a finite connected neighborhood structure $G=[P,N]$ with the maximal neighborhood degree ν_{max} of G. We denote by C_n the complete graph with n points. We assume that G is not the complete graph with ν_{max} points.

- Let $p_\varepsilon \in P$ be a point with the maximal neighborhood degree $m=\nu(p_\varepsilon)=\nu_{max}$. The subgraph which is built up by all points $q \in N(p_\varepsilon)$ of the point neighborhood of p_ε must be different from C_m with respect to the assumption, and it must be different from C_{m-1} because otherwise the point set $\{p_\varepsilon\} \cup N(p_m)$ leads to C_m in contradiction to our assumption.

- Therefore there are at least two points $p_1,p_2 \in N(p_\varepsilon)$ which are not neighbored to one another.

- If we delete the point p_ε and the edges outgoing from p_ε, we get a new graph G' with $\nu(q) < \nu_{max}$ for all $q \in N(p_\varepsilon)$.

- Let be $R=\{p_\varepsilon\}$ the set of all up to now deleted points, $S=P-R$ the set of all remaining points, and $k=\varepsilon$.

- Now, we select any point p_{k-1} with $\nu(p_{k-1}) < \nu_{max}$ which must be different from p_1 and p_2 (at least one point neighbored to R fulfills this condition). Then we build up the sets $R \cup \{p_{k-1}\} \rightarrow R$ and $S-\{p_{k-1}\} \rightarrow S$. This selection is continued up to $k=2$.

- The last two points p_2 and p_1 are not neighbored to one another. Therefore, they can be colored with the same color.

- The points $p_3,p_4,...p_{\varepsilon-1}$ can be colored successively by any of ν_{max} colors because each of these points has almost $\nu_{max}-1$ already colored neighbors.

- Finally all points are colored up to the last point p_ε. This point p_ε has ν_{max} already colored neighbors. But there are at least two points (namely p_1 and p_2) among these neighbors which have the same color. Therefore we have in $N(p_\varepsilon)$ at most $\nu_{max}-1$ colors, and p_ε can be colored with one of ν_{max} colors to guarantee a regular coloring of the whole graph.

The argumentation given here cannot be used for $\nu_{max}=2$ because for such a neighborhood structure the deconstruction of the neighborhood structure can only performed about p_1 and p_2 (however, these two points must be remained up to $k=2$). But $\nu_{max}=2$ describes either an unbranched tree or a circle. In both cases, we need at most three colors. And finally, we need for complete graphs C_n exactly n colors.

We see that the theory of neighborhood structures and the neighborhood method as graph searching can play a role also in graph theory: The common proof of the theorem of Brooks runs to several pages in graph theory textbooks [Sa70].

1.4 Dilatation and Erosion

1.4.1 Metric Spaces

This chapter shall begin with explanation of some notions which are essential for "continuous mathematics". The notions to be explained are not the basis of the further development of the theory. But they heuristically supply the understanding of the theory of digital image processing.

Def. 1.4.-1: A *metric space* is a set R of elements ("points" of the space) in which a metric is defined. That means that for any two elements $x,y \in R$ a non-negative real number $d(x,y)$ as *distance* between x and y is given which satisfies the following conditions:

$$d(p,p) = 0 \quad ,$$

$$d(p,q) = d(q,p) \quad ,$$

$$d(p,q) > 0 \quad \text{for} \quad p \neq q \quad ,$$

$$d(p,q) + d(q,r) \geq d(p,r) \quad .$$

This definition agrees completely with definition 1.2.-9. But in section 1.2.4, we have investigated graphs, especially finite graphs. It was out of our conception that there are "any infinitely small" distances. Now, in this section, a continuous set of points will be considered where two different points can lie infinitely near together.

In image processing, both conceptions are usual - the conception of discrete graphs and the conception of the continuous Euclidian space. Both mathematical models have their own meaning and significance in description of reality. On the one hand, we have to work with the discretly constructed computer memory. Otherwise, we can refer to the apparently continuous photographs. One can even argue that image formation by electro-magnetic waves is continuous. But is that so? Already in 1905, Einstein had introduced the corpuscular nature of light to explain the photo effect...

We will not support the sterile dispute whether images are "actually" discrete or continuous. Rather, each point of view shall be applied there where we reach better, faster, and more obvious results. The preponderance of discreteness in this book is a logical conclusion from the undervaluation of discreteness elsewhere.

Def. 1.4.-2: The set of all points $y \in R$ with a distance $d(y,x) < \delta$ in respect of a given point $x \in R$ for $\delta > 0$ is the *δ-surrounding* $S_\delta(x)$ of the point x.

Def. 1.4.-3: A point $a \in R$ is denoted as *limiting point of set* $M \subseteq R$ if each surrounding $S_\delta(a)$ contains an infinite number of points of M. The set of all limiting points of a set M will be symbolize by $H(M)$. A point $b \in R$ is a *contact point of set* $M \subseteq R$ if each surrounding $S_\delta(b)$ contains at least one point of M. The set of all contact points of a set M will be symbolize with $G(M)$. A point $m \in M \subseteq R$ is an *isolated point* of M if m is not a limiting point of M.

It is to be taken into consideration that each surrounding $S_\delta(m)$ always contains in respect to definition 1.4.-2 the point m. Therefore, the two notions of neighborhood and surrounding are not identical. In a neighborhood structure, there are no limiting points. But all points $p \in P$ of the neighborhood structure $[P,N]$ are contact points of P. This follows from definition 1.4.-3 because $G(M)=M \cup H(M)$.

Def. 1.4.-4: A set $M \subseteq R$ is *closed* if $M=G(M)$. The set M is *dense* if $M \subseteq H(M)$, and *perfect* if $M=H(M)$. We denote a set $M \subseteq R$ as *open* if its complement $R-M \subseteq R$ is closed. A perfect set is denoted as *continuum*.

These notions shall be illustrated by some examples:

- Given the set R of all real numbers with a metric defined by the distance function $d(x,y)=|x-y|$. The set Z of all integer numbers as subset of the set R of all real numbers consists only of isolated points. Because $Z=G(Z)$ and $H(Z)=\varnothing$, Z is closed.

- The set Y of all rational numbers $1/n$ with $n \in \{1,2,3,...\}$ as subset of the set R of all real numbers consists also only of isolated points. But it is $H(Y)=\{0\}$.

- Be X the set of all rational numbers as subset of the set R of all real numbers. Each point $x \in X$ is limiting point of X so that $X \subseteq H(X)$ follows. Therefore X is dense. But X is not closed because the irrational numbers do not belong to X although all irrational numbers are limiting points of X.

- Let U be the set of all numbers u with $0 \leq u \leq 1$. This set is closed and dense, and therefore U is a perfect set.

Def. 1.4.-5: A point $a \in M \subset R$ is denoted as *inner point* of M in respect to the metric space R if a is not contact point of the complementary set $R-M \subseteq R$, i.e. if there is a surrounding $S_\delta(a) \subset M$.

Def. 1.4.-6: A point $b \in M \subset R$ which is not an inner point of M shall be called *boundary point* of the set M.

Because of the definitions 1.4.-4/5, it follows that a set M is open if and only if all points of M are inner points of M. Therefore we conclude that open sets have no boundary points.

The interval ("open interval") of real numbers x with $0<x<1$ is an open set and has no boundary points. The segment ("closed interval") of real numbers x with $0 \leq x \leq 1$ has the both boundary points 0 and 1 because each surrounding of these two points contains also points which are outside of the segment.

Def. 1.4.-7: The set B of all points $r \in M \subseteq R$ with $S_\delta(r) \cap M \neq \emptyset$ and $S_\delta(r) \cap R-M \neq \emptyset$ is the *boundary* of the set M.

It must be emphasized that the common analysis (calculus) and the functional analysis are built up on these notions as well as the set-theoretical topology [Al56, Jä87, Qu79]. But these notions are constituent parts of a "think theory". A computer is - like a human being -not able to give explicitly the boundary points of the closed set $\{(x,y\} \mid x^2+y^2 \leq 1\}$, i.e. of the circular area in the Euclidian plane. And a computer is - like a human being - not able to work with the "exact value of π". Both - the computers and the human beings - use solely symbolically that $4 \cdot \pi/2$ is the same as $2 \cdot \pi$.

> "We should not forget that the mathematical formalism of modern scientific theories uses such mysterious artefacts as 'real numbers'." [Primas, Pr81]

Contrary to that, digital image processing as represented in this book is a "count theory". All investigated quantities can be determined by counting or by arithmetic expressions of counting results because we work always only with finite images. Now, a computer is directly destined for counting and arithmetic operations.

1.4.2 Boundaries and Cores in Neighborhood Structures

The fact that any formulation is "obvious" has its foundation in habit. According to the set-theoretical way of looking in the foregoing section, the closed sets $M=\{(x,y) \mid x^2+y^2 \leq 1\}$ and $M'=\{(x,y) \mid x^2+y^2 \geq 1\}$ both have the same boundary $B=\{(x,y) \mid x^2+y^2=1\}$ in the Euclidian plane. In this sense, we are in the habit of thinking that two sets are separated by their common boundary.

But we must not forget that the Euclidian geometry and the set-theoretical investigation of a metric space are only theoretical models. Look - for example -

to a real scene: A glass top lies on an iron cube. What is the boundary between glass and iron or between iron and air? Does each material have its own boundary? Does the notion "common boundary" reflect the reality?

It would be nonsense because of the difficulties arising with such questions to dispense with the whole "continuous mathematics". But it is of interest to look at the possible consequences of a discrete model in a long series of logical conclusions. Now, digital image processing enforces such an approach.

Like other situations, we have had predecessors already who were thinking about that problems of continuity and discreteness, respectively. One of them was the English mathematician Clifford (1845-1879). Clifford's interests were directed mainly to geometry. In his paper "The Postulates of the Science of Space", he has been investigated problems which are essential for us also today.

> "I have here a row of two different kinds of marbles, some white and some black. This aggregate is divided into two parts, as we formerly supposed the line to be. In the case of the line the boundary between the two parts is a point which is the element of which the line is an aggregate. In this case before us, a marble is the element; but here we cannot say that the boundary between the two parts is a marble. The boundary of the white parts is a white marble, and the boundary of the black parts is a black marble; these two adjacent parts have different boundaries.
>
> Similiarly, if instead of arranging my marbles in a series, I spread them out on a surface, I may have this aggregate divided into two portions - a white portion and a black portion; but the boundary of the white portion is a row of white marbles, and the boundary of the black portion is a row of black marbles.
>
> And leastly, if I made a heap of white marbles, and put black marbles on the top of them, I should have a discrete aggregate of three dimensions divided into two parts: the boundary of the white part would be a layer of white marbles, and the boundary of the black part would be a layer of black marbles. In all these cases of discrete aggregates, when they are divided into two parts, the two adjacent parts have different boundaries." [Clifford, Cl63]

Def. 1.4.-8: A point $p \in M$ of a subset $M \subseteq P$ of the neighborhood structure $[P,N]$ is called *inner point* or *core point* of M if $N(p) \subseteq M$. A point $q \in M$ which is not a core point of M is a *boundary point* of M.

Def. 1.4.-9: The set $K(M)$ of all core points of the set $M \subseteq P$ of the neighborhood structure $[P,N]$ is the *core* of M. The set $R(M)$ of all boundary points of M is denoted as *boundary* of M.

The empty set M contains neither boundary points nor core points so that $R(\emptyset)=K(\emptyset)=\emptyset$ follows. The set P of a neighborhood structure $[P,N]$ contains only core points because for each point $p\in P$ the relationship $N(p)\subseteq P$ is fulfilled. Therefore we have $R(P)=\emptyset$ and $K(P)=P$. For any sets $M\subseteq P$, it follows $M=R(M)\cup K(M)$ and $R(M)\cap K(M)=\emptyset$.

Using the both definitions 1.4.-8/9, some theorems can be proved [Vo88]:

Theorem 1.4.-1: Each nonempty subset $M\subset P$ of a connected neighborhood structure $[P,N]$ contains boundary points.

Theorem 1.4.-2: Each point $p\in K(M)$ of a proper subset $M\subset P$ of a connected neighborhood structure $[P,N]$ is connected in M with at least one point $q\in R(M)$.

Theorem 1.4.-3: If the boundary $R(M)$ of a nonempty proper subset $M\subset P$ of a connected neighborhood structure $[P,N]$ is connected, the set M is also connected.

Theorem 1.4.-4: The nonempty core $K(M)$ of a set $M\subset P$ of the neighborhood structure $[P,N]$ is separated from the complementary set $P-M\subset P$.

The connectivity of the neighborhood structures assumed for these theorems is very essential for their validity. It is shown in figure 1.4.-1 that there are in nonconnected neighborhood structures nonempty subsets $M\subset P$ without boundary points. And we can see that subsets are possible which are built up only by boundary points.

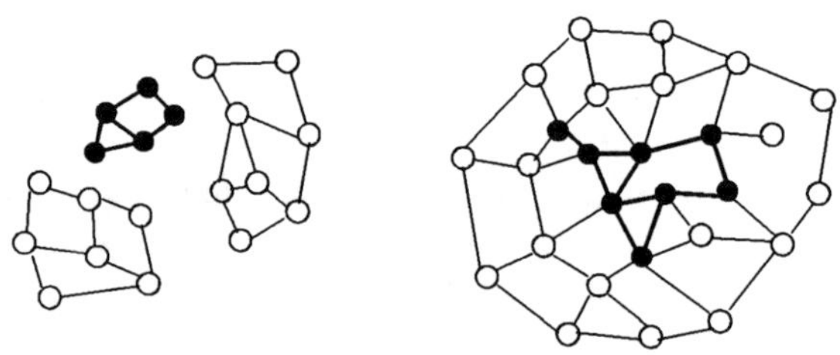

Fig. 1.4.-1: Subset $M=K(M)$ in a non-connected structure (left) and subset $M=R(M)$ in a structure (right)

Def. 1.4.-10: Let $M\subseteq P$ be a subset of a neighborhood structure $[P,N]$. All the points $q\in P-M\subset P$ with $N(q)\cap M\neq\emptyset$ are named *point neighbors* of M. The set $N(M)$ of all point neighbors of M is the so-called *point neighborhood* of M.

For boundary points of sets M in neighborhood structures, an analogous property to boundary points in metric spaces is given: the neighborhood of a boundary point or the surrounding of a boundary point contains points of $P-M\subset P$. The following general relationships are fulfilled for subsets $M\subseteq P$ of neighborhood structures $[P,N]$:

$$N(M) = R(\overline{M}) \quad , \quad R(M) = N(\overline{M}) \quad ,$$

$$N(M) \cap M = \emptyset \quad ,$$

$$\overline{N(M)} = M \cup K(\overline{M}) \quad , \quad N(K(M)) \subseteq R(M) \quad .$$

Because of the previous definitions to neighborhood structures, a new geometry follows - the geometry of neighborhood structures as investigated and developed especially by Hübler [Hü87, Hü87a, Hü89]. It is senseless to ask which geometry would be the "right" geometry. Each geometry is a consistent mathematical model, and it is the better the more appropriately it describes the parts of reality which are to be investigated.

The Euclidian geometry and the set-theoretical topology are only weakly qualified for description of Clifford's marble world or for description of the reality of digital images. But the theory of neighborhood structures - and therefore the theory of loopless undirected graphs without multiple edges - is in contrast a suitable model for digital images.

1.4.3 Set Operations and Set Operators

Digital image processing is not only a "static" science - that means a "picturology" which is built up on axioms and definitions and yields conclusions and theorems - but it is also a modern part of informatics with algorithms and methods in the centre of interest. An example for information processing has been given in section 1.3.3 with graph search. Determining the core $K(M)$ of a set M or the boundary $R(M)$ can be likewise in the centre of interest. This could be more important than determining any special properties of cores and boundaries. To emphasize the action of core determining or boundary formation, we can use *operators* which work on sets $M\subseteq P$ of a neighborhood structure $[P,N]$, and whose working can be symbolized by a core building operator $\boldsymbol{K}$ or a boundary creation operator $\boldsymbol{R}$. In this sense, we know the following operators up to now:

- $\boldsymbol{I}$ identity operator with $\boldsymbol{I}M=M$,
- $\boldsymbol{C}$ complement operator with $\boldsymbol{C}M=P-M$,
- $\boldsymbol{K}$ core building operator with $\boldsymbol{K}M=K(M)$,
- $\boldsymbol{R}$ boundary building operator with $\boldsymbol{R}M=R(M)$,
- $\boldsymbol{N}$ neighborhood operator with $\boldsymbol{N}M=N(M)$.

In these equations, the application of a set building operator to a set stands in the left, i.e. the operator is the symbolical description of an algorithm in respect to the given neighborhood relation. At the right side, the results of these operations are given.

Some of the above equations can be written as proper operator equations, namely $\boldsymbol{N}=\boldsymbol{RC}$ or $\boldsymbol{R}=\boldsymbol{NC}$. The properties and laws corresponding to the given operators are independent of the respective sets M. Building up the point neighborhood $N(M)$ of a set M or building up the boundary of the complementary set - in both cases the same result arises.

Therefore, the essence of these operations is contained in the processes rather in the sets to be investigated. Because an equation as $N(M)=\boldsymbol{NM}=\boldsymbol{RCM}=R(P-M)$ is valid for any sets, the explicit declaration of the respective set can be omitted. When we work with more than one operator, the operator sequence is determined by the set M on the right.

In the following operator table, the relationships are shown which are to be taken into account for successively performed operations, i.e. "operator multiplications". It is to see that we cannot give expressions for $\boldsymbol{KK}$ or some other operator products. These operators are not elements of an algebraic group (see section 5.2.2).

Table 1.4.-1: Operator table for operator multiplication

	2nd operator				
1st operator	**I**	**C**	**K**	**R**	**N**
I	**I**	**C**	**K**	**R**	**N**
C	**C**	**I**			
K	**K**			**Z**	**Z**
R	**R**	**N**		**R**	**N**
N	**N**	**R**			

The operator $\boldsymbol{Z}$ is the "zero operator" because the boundary $R(M)$ and the neighborhood $N(M)$ of a set M contain no core points. Therefore the operation sequences $K(R(M))$ and $K(N(M))$ yield always the empty set $\varnothing$ as result, i.e. $\boldsymbol{Z}M=\varnothing$.

1.4.4 Dilatation and Erosion

In this section, we will introduce some set operators important in image processing. But it is always to be taken into account that these set operators work on sets in neighborhood structures - in contrast to operations leading to union or intersection in unstructured sets.

Def. 1.4.-11: The set $M \cup N(M)$ in respect of a set $M \subset P$ of a neighborhood structure $[P,N]$ is denoted as *dilated* set $D(M)$. The set $D(M)$ is obtained by application of the *dilatation operator* $\boldsymbol{D}$ to M.

Def. 1.4.-12: The set $M-R(M)=K(M)$ in respect of a set $M \subset P$ of a neighborhood structure $[P,N]$ is denoted as *eroded* set $E(M)$. The set $E(M)$ is obtained by application of the *erosion operator* $\boldsymbol{E}$ to M.

The notions *dilatation* and *erosion* play an important role in the *Mathematical Morphology*. The discipline of Mathematical Morphology is the theoretical basis for digital image processing as developed and applicated by Serra and his school in Fontainebleau for some decades. The fundamental definitions are that of Minkowski addition and Minkowski subtraction of point sets - generally in metric spaces, especially in the Euclidian plane [Se84].

We will show here that the fundamental properties of dilatation and erosion are not restricted to set-theoretical disciplines but can be found and applied also in discrete structures. For dilatation, we obtain the relationship

$$D(M) = M \cup N(M) = \overline{\overline{M} \cap \overline{N(M)}} = \overline{\overline{M} \cap (M \cup K(\overline{M}))}$$
$$= \overline{(\overline{M} \cup M) \cup (\overline{M} \cap K(\overline{M}))} = \overline{\boldsymbol{K}(\overline{M})}$$

so that the operator equation $\boldsymbol{D}=\boldsymbol{CKC}=\boldsymbol{CEC}$ is fulfilled. Here, we have replaced the core building $\boldsymbol{K}M$ by the erosion $\boldsymbol{E}M$. The result of both operations is the same. But using $\boldsymbol{K}M$, we imagine that "it remains the core of set M" whereas using $\boldsymbol{E}M$, we think that "the boundary of set M is removed".

Because always $C(C(M))=M$ or $\boldsymbol{CC}M=M$ is valid, we can derive from $\boldsymbol{D}=\boldsymbol{CEC}$ by left and right multiplication with $\boldsymbol{C}$ that $\boldsymbol{CDC}=\boldsymbol{E}$.

Formula 1.4.-1:

$$\boldsymbol{CEC} = \boldsymbol{D} \quad , \quad \boldsymbol{CDC} = \boldsymbol{E}$$
$$\boldsymbol{EC} = \boldsymbol{CD} \quad , \quad \boldsymbol{CE} = \boldsymbol{DC} \quad .$$

Contrary to the operator $\boldsymbol{C}$, the operators $\boldsymbol{D}$ and $\boldsymbol{E}$ have no inverse operators, i.e. it is not possible in general to undo the operations dilatation and erosion. The reason is that a one-to-one mapping can be realized neither by $\boldsymbol{D}$ nor by $\boldsymbol{E}$ from the set P^* of all subsets of P into P^* (see figures 1.4.-2/3). But if we found relationships like $\boldsymbol{CDC}=\boldsymbol{E}$ then we will denote the two operators $\boldsymbol{D}$ and $\boldsymbol{E}$ as *dual* to one another.

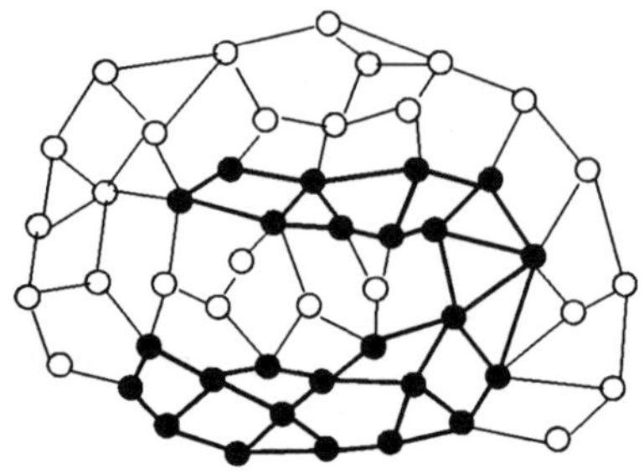

Fig. 1.4.-2: There is no inverse operation of the dilatation of a set M (points ●)

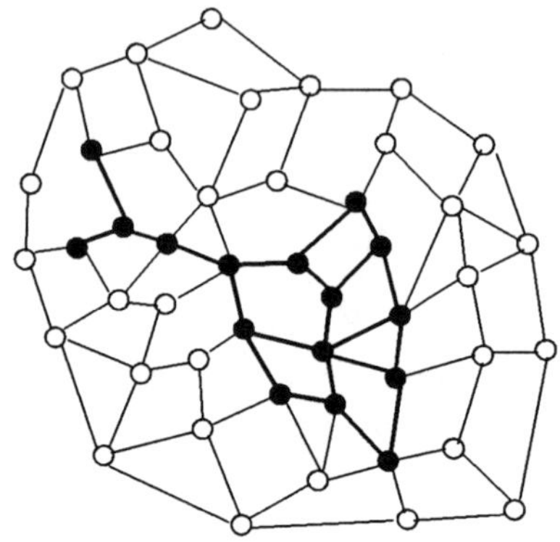

Fig. 1.4.-3: There is no inverse operation of the erosion of a set M (points ●)

The dilated set $D(M)$ contains the set M as subset, and it is $D(D(M)) \supseteq D(M)$ etc. With respect to erosion, the set $E(M)$ is always a subset of M. Therefore, we obtain for each proper nonempty subset $M \subset P$ the inclusion sequence

$$\emptyset \subset \ldots \subset \boldsymbol{E}^2 M \subset \boldsymbol{E} M \subset M \subset \boldsymbol{D} M \subset \boldsymbol{D}^2 M \subset \ldots \subset P$$

for finite neighborhood structures $[P,N]$.

1.4.5 Opening and Closing

It has been shown in section 1.4.4 that the operators $\boldsymbol{D}$ and $\boldsymbol{E}$ possess no inverse operators. Nevertheless, an erosion can be inverted "approximately" by a dilatation, and a dilatation can be inverted "nearly" by an erosion. The operators $\boldsymbol{D}$ and $\boldsymbol{E}$ do not negate each other logically but the expressions $\boldsymbol{DE}$ or $\boldsymbol{ED}$ yield a new quality in the sense of dialectical negation of negation.

In the figures 1.4.-2/3, the new quality is to seen because by $E(D(M))$ small "gaps" in a set M are closed, and that by the combined operation $D(E(M))$ narrow "bridges" of a set can be destroyed. The operator products $\boldsymbol{DE}$ or $\boldsymbol{ED}$ enable so to speak an "image enhancement". But using this notion, it is tacitly assumed that one knows what is an "enhanced" image or a "better" image: An image scientists who have to count blobs or fibres in figure 1.4.-4 will not be satisfied with the result yielded by $\boldsymbol{DE}$ or $\boldsymbol{ED}$.

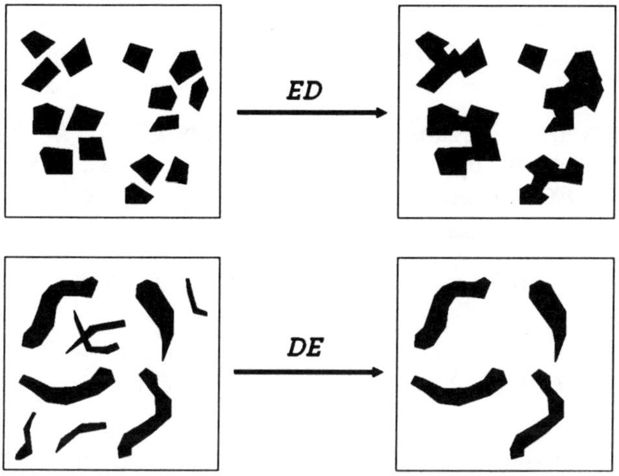

Fig. 1.4.-4: In these cases, the operators ***ED*** and ***DE*** give no "enhancement" of the image

But in the next figure 1.4.-5, we see that a "better" image can be reached because noise (single black and white points) and perturbations of the image are removed.

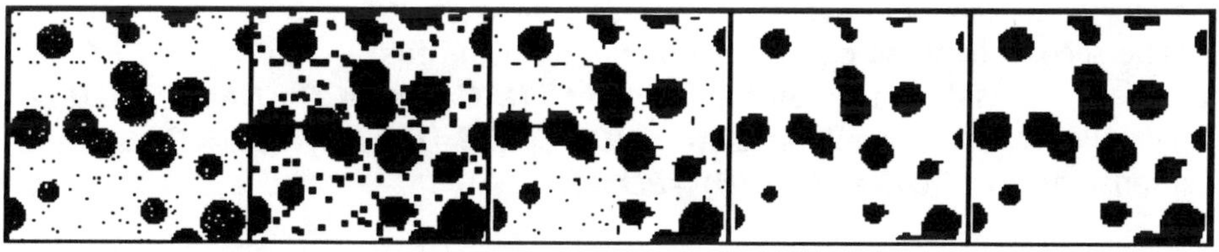

Fig. 1.4.-5: Image enhancement by application of ***D***, ***S***=***ED***, ***ES***=***EED***, and ***OS***=***DEED***

Def. 1.4.-13: The operation $S(M)=E(D(M))$ is named *closing*, and the operator $\boldsymbol{S}=\boldsymbol{ED}$ is the *closing operator.*

Def. 1.4.-14: The operation $O(M)=D(E(M))$ is named *opening*, and the operator $\boldsymbol{O}=\boldsymbol{DE}$ is the *opening operator.*

Generally, it is $S(M)\neq O(M)$. This difference between operators $\boldsymbol{S}$ and $\boldsymbol{O}$ can more exactly described by means of relationship $N(K(M))\subseteq R(M)$ by

$$O(M) = D(E(M)) = E(M) \cup N(E(M))$$
$$= K(M) \cup N(K(M)) \subseteq K(M) \cup R(M) = M \quad .$$

Because $O(M)$ is always a subset of M, we obtain $\overline{O(M)} \supseteq \overline{M}$ for the complementary set. Further, we have ***CO*=*CDE*=*ECE*=*EDC*=*SC*** so that in respect to the duality ***CD*=*EC*** between ***D*** and ***E*** the duality ***CO*=*SC*** between ***O*** and ***S*** follows. Therefore, we obtain

$$\boldsymbol{CO}M = \boldsymbol{SC}M = \boldsymbol{S}\overline{M} \supseteq \overline{M}$$

or

$$O(M) = D(E(M)) \subseteq M \subseteq E(D(M)) = S(M) \quad .$$

Using more dilatation and erosion operators, we get further operator equations. Let $M'=E(M)$ be a set of points of the neighborhood structure $[P,N]$. It is not possible to choose any M' because M' must be the result of an erosion operation. Because of $E(M)=K(M)$ and $N(K(M))\subseteq R(M)$, the neighborhood of M' is a subset of the boundary $R(M)$. Each point of $N(K(M))=N(E(M))$ possesses neighbors in the complementary set $P-M\subseteq P$ and in $K(M)$ if $K(M)\neq\emptyset$. Therefore we obtain $R(K(M)\cup N(K(M)))=R(D(K(M)))=N(K(M))$. By erosion of set $D(K(M))$, we get once more the set $K(M)$. From $E(D(E(M)))=E(M)$, the operator equations ***EDE*=*E*** and ***DED*=*D*** follows.

The second of these equations follows from the first by ***CEDEC*=*CEC*=*D*** because the left side can be reformulated to ***CEDEC*=*DCDCD*=*DECCD*=*DED***. Especially we find the relationships ***OO*=*DEDE*=*D*(*EDE*)=*DE*=*O*** and the relationships ***SS*=*EDED*=*E*(*DED*)=*ED*=*S***.

<u>Formula 1.4.-2:</u>

$$\boldsymbol{CO} = \boldsymbol{SC} \quad , \quad \boldsymbol{OC} = \boldsymbol{CS} \quad ,$$
$$\boldsymbol{SE} = \boldsymbol{E} = \boldsymbol{EO} \quad ,$$
$$\boldsymbol{DS} = \boldsymbol{D} = \boldsymbol{OD} \quad ,$$
$$\boldsymbol{OO} = \boldsymbol{O} \quad , \quad \boldsymbol{SS} = \boldsymbol{S} \quad .$$

It is meaningful for digital image processing that the multiple application of opening or closing operators yields no further changing and therefore no further enhancement of the visual impression of the image. To close larger gaps, we have to work with multiple dilatation operators and multiple erosion operators. So we obtain operator expressions such as ***EEDD*** or ***DDEE*** which cannot be further simplified. Such operators are closing or opening operators of second step, and finally we have operators of higher order like $\boldsymbol{E}^n\boldsymbol{D}^n$ and $\boldsymbol{D}^n\boldsymbol{E}^n$ in general.

Maybe an important point of view is lost because of all the foregoing mathematics: Erosion and dilatation, opening and closing, building complementary sets and boundary sets etc. are image processing operations which have to run on the computer. The above operator equations express something about the properties of methods and procedures. Instead of working with seven successively called procedures

```
CALL    COMPLEMENT
CALL    CLOSING
CALL    COMPLEMENT
CALL    ERODE
CALL    COMPLEMENT
CALL    ERODE
CALL    COMPLEMENT
```

corresponding to the operator product ***CECECSC***, we can replace the procedure sequence by

```
CALL    OPENING
```

because

$$\boldsymbol{CE}(\boldsymbol{CEC})\boldsymbol{SC} = \boldsymbol{CEDSC} = \boldsymbol{C}(\boldsymbol{ED})\boldsymbol{SC} = \boldsymbol{CSSC} = \boldsymbol{CSC} = \boldsymbol{O} \quad .$$

Therefore, the theoretical investigation of algorithms and their properties can lead to faster and more effective procedures and programs. That is the significance of the things developed in this chapter.

2 Incidence Structures

2.1 Homogeneous Incidence Structures

2.1.1 Topological Problems

Topology is a branch of modern mathematics which deals mainly with infinite (not countable) continuous sets, i.e. with point sets where infinitely many points lie in the surrounding of a point.

The point sets investigated in image processing are of another type: They are almost finite sets, and the neighborhood of a point contains only a finite number of other points. In the sense of the set-theoretical topology, point sets of image processing are discrete point sets, i.e. sets where each point is an isolated point (see section 1.4.1).

Searching in a textbook of topology for the notion "discrete point set", one finds solely a hint to a structure consisting only of isolated points in which no neighborhoods, no connections, and no boundaries exist. These structures possess only a *discrete topology* [Ri75, Jä87].

The set-theoretical topology was introduced in 1914 by the book "Grundzüge der Mengenlehre" written by F.Hausdorff [Ha14]. The set-theoretical topology has in its abstract part the task to establish axiomatically the topological notion of space and to build up on this base a deductive theory:

> A *topological space* $[R,T]$ consists of a set R of space points and of a set T of subsets of R - so-called "open sets" - with following properties:
>
> 1. Any unions of open sets are open.
> 2. The intersection of two open sets is open.
> 3. The sets R and $\emptyset$ are open.

Together with the definitions of section 1.4.1, this definition yields the basis of set-theoretical topology. The topology is a widely developed mathematical discipline and very important for many modern branches of mathematics. But how far - unfortunately - the problems of topology are from the problems of image processing is shown in each textbook of topology.

In another part of topology, in the *combinatorical topology*, algebraic methods stand in the foreground of interest. The basis for this part is the notion of a *complex*. The axiomatically established investigation of combinatorical structure and combinatorical properties of complexes and the application of the theory of polyhedrons is the content of combinatorical topology [Ei52, Le52, Po56]. The complexes of combinatorical topology are the most important examples for discrete spaces. The set theory is dealing with zero-dimensional complexes, and one-dimensional complexes are investigated by graph theory.

What we need for image processing is a combinatorical topology in the sense that notions like neighborhood and connectivity can be exactly defined for elements of a discrete space. The direction of investigation is given in part 1 of this book (neighborhood structures, neighborhood graphs). But essential questions are open up to now which shall be investigated in this part:

- In a digital image, each point has the same number of neighbors. What laws can be found for regular graphs (or more general: for homogeneous structures)?

- Up to now, we have investigated only sets (zero-dimensional complexes) and graphs (one-dimensional complexes). What laws can be found in generalized higher dimensional complexes?

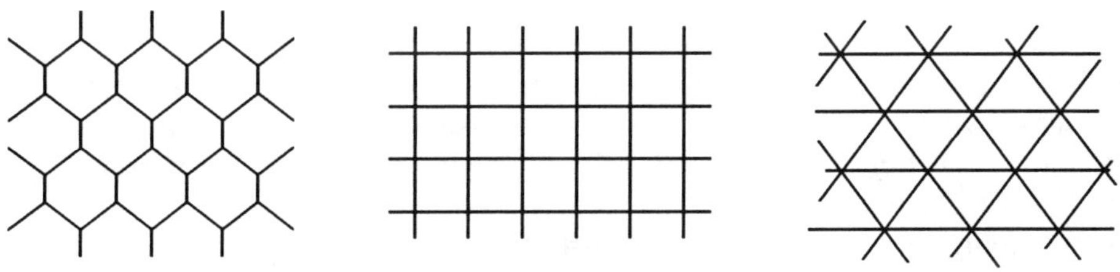

Fig. 2.1.-1: The three two-dimensional homogeneous structures $\Gamma(\nu,\lambda)$ with $(\nu,\lambda)=(3,6)$, $(4,4)$, and $(6,3)$ for two-dimensional image processing

In a series of foregoing papers, an attempt has been made to answer these questions in the following way [Vo86, Vo86a, Kl87, Kl87a, Vo87, Vo88]:

- Neighborhood structures $[P,N]$ are introduced as loopless undirected graphs without multiple edges.

- The neighbors $q\epsilon N(p)$ of each point $p\epsilon P$ are ordered cyclically as $<q_1,q_2,\ldots,q_n>$. Because of this ordering, we imprint an orientation on the neighborhood structure. Each directed edge (p,q) defines unambiguously a closed path in P which is denoted as *mesh* or as *directed way*. Points, edges and meshes represent an *oriented* neighborhood structure.

- Assuming that all points $p\epsilon P$ have the same number $\nu=\nu(p)$ of neighbors and that all meshes m have the same length $\lambda=\lambda(m)$, the neighborhood structure is a *homogeneous* neighborhood structure.

- The requirements ν=const and λ=const yield only then homogeneous structures of any size if $1/\nu+1/\lambda=1/2$ is fulfilled. The three homogeneous oriented neighborhood structures shown in figure 2.1.-1 are the only solutions of that diophantic equation. They are also the only possible homogeneous structures for two-dimensional image processing.

Each attempt to generalize this approach to a theoretical foundation of image processing - at least to enable description and investigation of three-dimensional structures - is miscarried in the past. But the three main formulas of two-dimensional oriented homogeneous neighborhood structures, namely

$$\sum_{p\in P} \nu(p) = \nu\,\varepsilon = 2\,\varkappa\ ,$$

$$\sum_{m\in M} \lambda(m) = \lambda\,\omega = 2\,\varkappa\ ,$$

$$\varepsilon - \varkappa + \omega = 0$$

with ε as point number, $\varkappa$ as edge number, and ω as number of directed ways (meshes) were so fruitful for image processing that the aim then was to somehow find a generalization. Finally, I found that the theory of complexes gives the right hints for continued conclusions. The systematically developed theory going beyond the first attempts [Le86, Ko87, Vo90, Vo91] shall be represented in the following chapters.

Independently from us, the basic idea was formulated also by Rosenfeld and his co-workers:

> "There is another method of representing a digital image by continuous image...Instead of taking a voxel to represent a digital point, we take a point, namely the same grid point p in R^3. We also take every unit grid line segment with its two end points in S, every unit square with four corner points in a 2·2 block of points of S, and every unit cube with eight corner points in a 2·2·2 block of points of S..." [Lee and Rosenfeld, Le85a]

2.1.2 Cellular Complexes

The theory of cellular complexes does not understand edges like the set-theoretical topology as consisting of infinite many points and also not like the theory of neighborhood structures as ordered or unordered point pairs [Ri75, He83, Ja83, Kl83, Le86, Ko87, Ko89a, Wi91]. Analoguously, neither the small squares of an orthogonal lattice are areas consisting of infinitely many points nor meshes in neighborhood structures are cyclically ordered point sets. There are only points, edges, and meshes which are objects of different kinds.

When we name points, edges, and meshes as objects of zero kind, first kind, and second kind, then a generalization is easily possible.

> "We consider R^3 as the union of the cubes (also called 3-cells or voxels) whose vertices (also called 0-cells) are integer points and edges (also called 1-cells) have length 1. The faces of the cubes are also called 2-cells. If two cells are distinct, we say that they are incident if and only if one is a subset of the other, and that they are *n*-adjacent if and only if there is an *n*-cell which is incident to both..." [Herman and Webster, He83].

Def. 2.1.-1: A *cellular complex* (or *complex*) $C=[E,B]$ is given by a set E of elements, the so-called *cells* of C, and by a binary, irreflexive, anti-symmetrical, and transitive *boundary relation* B. A non-negative integer number $dim(e)$ is attached to each cell $e \in E$ which is denoted as *dimension* of e. It is necessary for $(e,e') \in B \subset E \times E$ that $dim(e) < dim(e')$.

Def. 2.1.-2: The maximal value of all $dim(e)$-values is the *dimension* of the cellular complex. A cellular complex is locally finite if the following requirements are fulfilled:

- From $dim(e) > 0$, the existence of at least one $e' \in E$ follows with $(e',e) \in B$.
- From $(e,e') \in B$ and $dim(e')-dim(e) > 1$, the existence of at least one $e'' \in E$ follows so that $(e,e'') \in B$ and $(e'',e') \in B$.
- There are only a finite number of elements $e' \in E$ with $(e',e) \in B$ for each $e \in E$, i.e. with $(e,e') \in B$.

Corresponding to these definitions, a zero-dimensional complex contains only zero-dimensional elements which we will denote as *points*. With respect to definition 2.1.-1, the relation B is in this case an empty set because for all $e,e' \in E$ the equation $dim(e)=dim(e')=0$ is fulfilled. Therefore, a zero-dimensional complex $C=[E,\varnothing]$ is nothing more than a somewhat complicatedly described set E, and the theory of zero-dimensional complexes corresponds to the set theory.

A one-dimensional complex contains zero-dimensional and one-dimensional elements which shall be shortly denoted as *points* and *edges*. The set E is composed of the both disjoint sets E_0 of points and E_1 of edges. Because of definition 2.1.-1, we have $B \subseteq E_0 \times E_1$, and from definition 2.1.-2 follows that there is for each edge $e \in E_1$ at least one point $e' \in E_0$ with $(e',e) \in B$.

One-dimensional complexes of such universality can be described only by a very general graph theory: There are points, and there are edges. Points can be connected with edges by the relation *B*. But edges must be connected with at least one point.

However, what is it for a graph theory where an edge can be connected with more than two points? By the additional requirements that there are exactly two points $e', e'' \in E_0$ for each edge $e \in E_1$, and that there is maximally one edge $e \in E_1$ for two points $e', e'' \in E_0$, then such complex corresponds exactly to a neighborhood graph. Therefore, part 1 of this book contains the theory of those special one-dimensional complexes. For two-dimensional complexes, one can derive from the definitions 2.1.-1/2 that

$$B \subseteq (E_0 \times E_1) \cup (E_0 \times E_2) \cup (E_1 \times E_2)$$

is fulfilled. By a suitable specialization of two-dimensional complexes, the oriented neighborhood structures arise which are characterized by a point set E_0, an edge set E_1, and a mesh set E_2 [Vo86, Vo86a, Kl87, Kl87a, Vo87, Vo88].

Contrary to [Vo88], the detailed theory of oriented neighborhoods shall be omitted here. Instead of that, we will formulate some restrictions to the general definitions 2.1.-1/2 and develop a theory of structures which contains the elementary set theory, the theory of neighborhood graphs, and the theory of oriented neighborhood structures but also opens the way to higher-dimensional structures. Models related to theory of cellular complexes are known also in other fields as for instance in computer graphics [Br80, An85, Ed87, Li91].

> "In recent years, numerous topological models have been developed for boundary representation, and there have been important developments with respect to dimension and orientability ... 'Incidence graphs' are graphs or hypergraphs, where the nodes generally represent the cells of the modelled subdivision (vertex, edge, face etc.), and the edges represent the adjacency and incidence relations between these cells... these incidence graphs represent a finite collection of open disjoint *iD* cells ($0 \leq i \leq n$) whose union is an *nD* manifold. The boundary of each *iD* cell is equal to the union of cells which have a dimension lower than *i*. Each *iD* cell belongs to the boundary of an $(i+1)$-dimensional cell." [Lienhardt, Li91]

Edelsbrunner describes an analoguous approach of implementation where the incidence relations are represented by means of lists [Ed87]: Two lists of cells are associated to each *i*-dimensional cell e_i, a list of all $(i\text{-}1)$-dimensional cells incident to e_i and a list of all $(i+1)$-dimensional cells incident to e_i.

Contrary to the method developed here, the inhomogeneity of objects to be constructed in computer graphics requires rather the use of lists instead of the use of data arrays.

A special type of incidence graph was used in [An85] for description of topology of orientable closed surfaces. There is a "hypergraph" defined by a set of "nodes", a set of "arcs", and a set of "hyperarcs". Each node correponds to a face of the polyhedral surface. The arcs correspond to edges of the surface and join two nodes because an edge is incident with two faces. A hyperarc corresponds to a vertex of the surface and joins all nodes representing the faces which are incident to the vertex. Hypergraphs are therefore two-dimensional incidence structures dual to our oriented neighborhood structures.

The method of incidence graphs can be compared with Kovalevski's method of cellular complexes [Ko87, Ko89a]. Kovalevski shows that the method of cellular complexes gives fruitful tools for description of inhomogeneous discrete structures. He speaks of *abstract cell complexes*. A two-dimensional image is regarded as a mapping of a two-dimensional complex into a set of integers representing the gray values.

> "After having applied a segmentation procedure one obtains an image with comparatively large regions each having a constant gray value. A region may be considered as a 2-dimensional *block of cells*. The boundary of a region is a 1-complex which may have *branching points* at locations where three or more regions meet. The branching points are declared to be the 0-dimensional blocks. Parts of a boundary separated by branching points are the 1-dimensional blocks or *lines*...
>
> There exists a bounding relation between the blocks: lines may be bounded by some branching points, regions are bounded by lines and possibly by branching points. Thus the collection of blocks with the bounding relation may be regarded as a new abstract cell complex whose 2-cells are the regions, 1-cells are the lines and 0-cells are the branching points. We shall call this structure the *block complex* and its elements the *block cells*.
>
> The structure may be described on a computer by a list of block cells with pointers which represent the bounding relation while indicating, for each block cell, which other cells are bounding it and/or are bounded by it. This *topological* part of the structure may be regarded as a generalization of the well-known region adjacency graph...All these data compose the so-called *cell list*...The cell lists...make any topological and geometrical analysis of images efficient and simple. Adjacent regions, regions contained in each other, lines and points incident with each other or with some regions may be directly and quickly found in the list." [Kovalevski, Ko90]

2.1.3 Incidence Structures

As already mentioned in 1983 by Klette, the theory of discrete spaces useful in image processing can be developed using two basic notions [Kl83a]. At first, we have the theory of n-dimensional cellular spaces as discussed in the previous sections. The n-dimensional cells are defined by the n-dimensional unit cubes which build up the space. This approach is used also in the cuberille method (see section 3.1.1).

The second notion is that of the n-dimensional grid point space or the n-dimensional cubic lattice. Here, the lattice points are the basic elements of the theory. We will prefer this approach because the number theory offers many results which are useful or at least of interest for image processing problems.

But both approaches are possible ways to build up a theory of image processing. They are dual theories in any sense as the following table shows:

	cellular space C^3	grid point space Z^3
basic elements	unit cubes	lattice points
boundaries of basic elements	faces	edges
boundaries of boundaries	edges	faces
constructed elements	points	cubes

Def. 2.1.-3: An *incidence structure* (simpler denoted as *structure*) $\Sigma=[E,I]$ is given by a set $E=\bigcup_{i=0}^{n} E_i$ of elements and a binary, reflexive, symmetrical, and intransitive *incidence relation* $I\subset \bigcup_{i,k} E_i \times E_k$. The sets E_i are pairwise disjoint. The elements of set E_i are called *i-dimensional*. The number n is the *dimension* of the incidence structure.

The essential difference between an incidence structure and a cellular complex is the reflexivity, symmetry, and intransitivity of the incidence relation I on the one side and the irreflexivity, anti-symmetry, and transitivity of the boundary relation B on the other side. The reflexivity of the incidence relation I distinguishes an incidence structure also from neighborhood structures where the neighborhood relation is irreflexive. It shall be pointed out that the notion of incidence structure has sometimes also a restricted meaning [Be82].

Def. 2.1.-4: In a finite incidence structure, there are non-negative numbers $b_{kl}(e)$ for each k-dimensional element $e \in E_k$ which describe the number of l-dimensional elements $e' \in E_l$ with $(e,e') \in I$. The numbers b_{kl} are named *structure constants*.

In figure 2.1.-2, the relationships between k-dimensional elements and l-dimensional elements given by the incidence relation I are represented. The line between a k-dimensional element $e^{(k)}$ and an l-dimensional element $e^{(l)}$ symbolizes that $(e^{(k)},e^{(l)})\in I$.

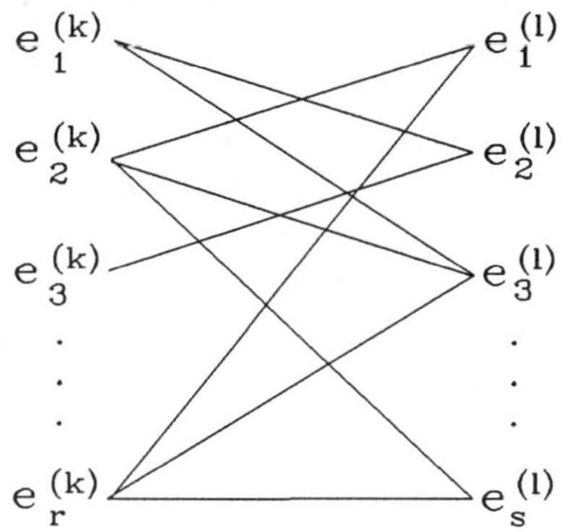

Fig. 2.1.-2: Relationships between k-dimensional elements and l-dimensional elements

This representation is useful also for $k=l$. But it must be taken into account that there are loop-lines from each element $e^{(k)}$ to itself because of the reflexivity of relation I. Therefore, we set always $b_{kk}=1$. Counting lines in figure 2.1.-2, we obtain

$$\sum_{e\in E_k} b_{kl}(e) = \sum_{e'\in E_l} b_{lk}(e')$$

for any k and l with $0\leq k,l\leq n$ with n as dimension of the incidence structure. This formula is named *matching theorem*. In respect to image processing, this is the most important formula for all n-dimensional incidence structures. In [Be82], this formula is denoted as "principle of double counting", and it is used for general one-dimensional incidence structures which are called "designs".

For one-dimensional neighborhood structures, the node theorem 1.1.-1

$$\sum_{p\in P} \nu(p) = \sum_{e\in E_0} b_{01}(e) = \sum_{e'\in E_1} b_{10}(e') = 2\varkappa$$

follows with $\nu(p)$ as number of directed edges outgoing from point p, and $\varkappa$ as number of all undirected edges.

2.1.4 Homogeneous Incidence Structures

As in the investigation of neighborhood structures, we get essential new insights if we restrict ourselves to homogeneous structures. A homogeneous incidence structure is characterized by $b_{kl}(e)=b_{kl}$ for all elements $e \in E_k$. Examples for homogeneous incidence structures are the regular graphs mentioned in section 1.1.3 and the two-dimensional structures shown in figure 2.1.-1.

Def. 2.1.-5: An incidence structure is called *homogeneous* if $b_{kl}(e)=b_{kl}=\text{const}$ for all elements $e \in E_k$ and all k,l with $0 \leq k,l \leq n$.

For finite n-dimensional incidence structures, $a_k=card(E_k)$ is the number of k-dimensional elements of the structure. Then, we obtain from the matching theorem the *matching formula*

Formula 2.1.-1:

$$a_k b_{kl} = a_l b_{lk} \qquad \text{for} \qquad 0 \leq k,l \leq n \; .$$

Given the number a_0 of points, then we obtain $a_k=a_0 b_{0k}/b_{k0}$. Because a_0 and a_k must be integer numbers, not all integers a_0 are allowed in general. Therefore the above relationships are a nonlinear diophantic equation system.

Using the similar formulas $a_k=a_m b_{mk}/b_{km}$ and $a_l=a_m b_{ml}/b_{lm}$, it follows from formula 2.1.-1 the relationship $a_m b_{mk} b_{kl}/b_{km}= a_m b_{ml} b_{lk}/b_{lm}$ so that we get with $a_m>0$ the equations

$$\frac{b_{mk}\, b_{kl}}{b_{lk}\, b_{km}} = \frac{b_{ml}}{b_{lm}} \qquad \text{for} \qquad 0 \leq k,l,m \leq n \; .$$

Like in the case of cellular complexes, we guarantee the conditions $a_m>0$ by the following axiomatic requirements:

- From $e \in E_k$ and $k>0$, the existence of at least one $e' \in E_l$ follows with $l<k$ and $(e,e') \in I$.

- From $e \in E_k$, $e' \in E_l$, and $k-l>1$, the existence of at least one $e'' \in E_m$ follows with $k<m<l$ and $(e_k,e_m),(e_m,e_l) \in I$.

The above formulas and the generalizations

$$\frac{b_{mr}\, b_{rs} \cdots b_{tl}}{b_{lt} \cdots b_{sr}\, b_{rm}} = \frac{b_{ml}}{b_{lm}}$$

are combinatorical laws for the structure constants which are based only on the matching theorem and on the requirement of homogeneity.

It cannot be the aim of this book to build up a full mathematical theory of incidence structures. But it is certainly meaningful to represent explicitly some structures to illustrate the notion of a structure and to give the certainty that there are incidence structures in fact.

(2,1,1,1,2,1)

(c,c,1,2,2,1)

(4,6,1,3,2,1)

(5,10,1,4,2,1)

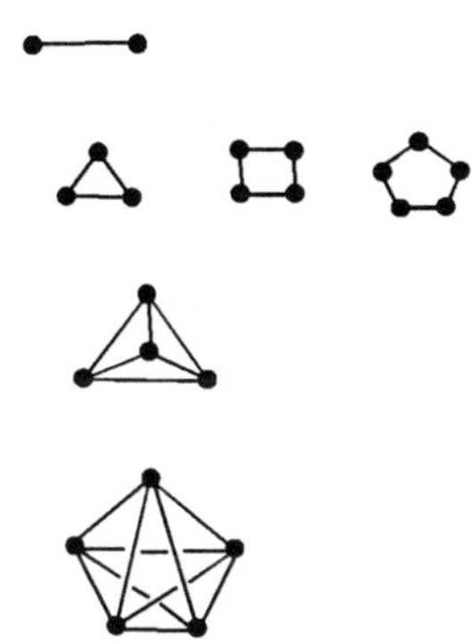

Fig. 2.1.-3: One-dimensional incidence structures

One-dimensional incidence structures are represented in figure 2.1.-3. We will characterize such structures by tupels $(a_0,a_1, b_{00},b_{01},b_{10},b_{11})$. The matching formula requires that $a_0b_{01}=a_1b_{10}$ or $\varepsilon\cdot\nu=\varkappa\cdot 2$ if we assume $b_{10}=2$ because of geometrical reasons. Now, if the tupel $(\varepsilon,\nu,\varkappa)$ is a solution of this diophantic equation then $(c\varepsilon,\nu,c\varkappa)$ and $(\varepsilon,c\nu,c\varkappa)$ with $c>0$ are solutions, too. In the first case, we get a c-fold repetition of the structure $(\varepsilon,\nu,\varkappa)$ what is only a senseless and tedious construction. In the second case, we get graphs with multiple edges - a solution without interest for theory of neighborhood graphs and image processing.

The complete graphs $(\varepsilon,\nu,\varkappa)=(c,c-1,c(c-1)/2)$ are also incidence structures. But we will not search here for all possible solutions. It is only important for image processing that the homogeneous incidence structures $(\varepsilon,\nu,\varkappa)=(c,2,c)$ with any large values of c are the base for one-dimensional image processing (see chapter 5.1).

Two-dimensional incidence structures are characterized by a 12-tupel of numbers a_0,a_1,a_2 and $b_{00},b_{01},b_{02},b_{10},b_{11},b_{12},b_{20},b_{21},b_{22}$. In figure 2.1.-4, these numbers are given for two examples - an octahedron and a cube (hexahedron).

In the left of figure 2.1.-4, pseudo-three-dimensional representations of these polyhedrons are shown where the polygonal sides of that polyhedrons are the two-dimensional elements of the incidence structures. The right diagrams show

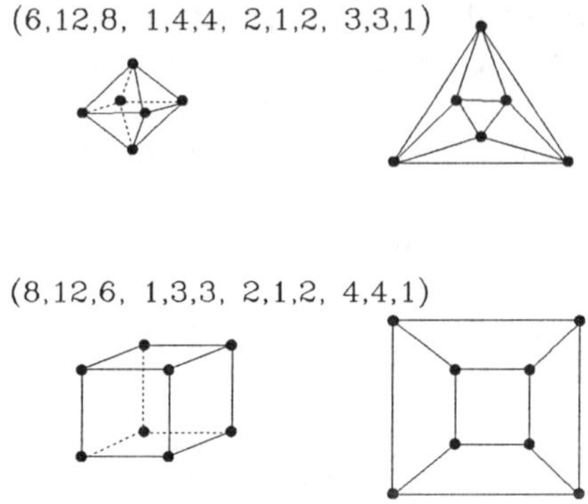

Fig. 2.1.-4: Two-dimensional incidence structures

the corresponding planar oriented neighborhood structures where the meshes are the two-dimensional elements.

Using the geometrically motivated assumptions $b_{10}=b_{12}=2$, $b_{01}=b_{02}$, $b_{20}=b_{21}$, $b_{00}=b_{11}=b_{22}=1$, we can characterize all two-dimensional incidence structures by 12 numbers $(a_0,a_1,a_2,b_{00},\ldots b_{22})=(\varepsilon,\varkappa,\omega,1,\nu,\nu,2,1,2,\lambda,\lambda,1)$. By means of the five quantities $\varepsilon,\varkappa,\omega,\nu$, and λ, we obtain three relationships given by the matching formula 2.1.-1:

$$a_0 b_{01} = a_1 b_{10} \quad \rightarrow \quad \varepsilon\cdot\nu = \varkappa\cdot 2 \;,$$

$$a_0 b_{02} = a_2 b_{20} \quad \rightarrow \quad \varepsilon\cdot\nu = \omega\cdot\lambda \;,$$

$$a_1 b_{12} = a_2 b_{21} \quad \rightarrow \quad \varkappa\cdot 2 = \omega\cdot\lambda \;.$$

There are a lot of two-dimensional incidence structures of this kind (see [Vo88]). But like to the one-dimensional case, only few of these structures are suited for image processing, namely the toroidally closed lattices and the infinitely spreaded nets of figure 2.1.-1.

Finally, it shall be shown that three-dimensional incidence structures also exist. Such structures are mainly characterized by four numbers $(a_0,a_1,a_2,a_3)=(\varepsilon,\varkappa,\mu,\zeta)$ of points, edges, meshes, and cells. Further, we have 16 structure constants and 6 matching relationships between them. Because we will investigate these relationships in the following sections and chapters in general, here only a simple example is given for demonstration (figure 2.1.-5). The structure constants of this example are

$$(b_{00},b_{01},b_{02},b_{03})=(1,4,6,4), \quad (b_{10},b_{11},b_{12},b_{13})=(2,1,3,3),$$

$$(b_{20},b_{21},b_{22},b_{23})=(4,4,1,2), \quad (b_{30},b_{31},b_{32},b_{33})=(8,12,6,1).$$

This incidence structure contains $a_0=\varepsilon=16$ points, $a_1=\varkappa=32$ edges, $a_2=\mu=24$ meshes as for instance $<a,b,c,d>$ or $<a,b,f,e>$, and $a_3=\zeta=8$ cells like that built up by points a,b,c,d, e,f,g,h. Thereby the "out lying" cell is also to count, similarly to the "out lying" meshes in the case of the right diagrams of figure 2.1.-4.

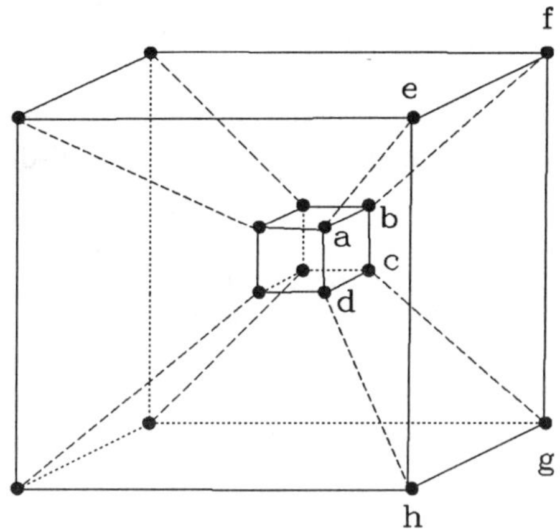

Fig. 2.1.-5: Three-dimensional incidence structure

The object in figure 2.1.-5 represents one of the six regular bodys in the four-dimensional Euclidean space as known since the last century [Sc84, Po04] (for $n>4$, there are always only three regular convex bodies in E^n [Bö91]). The regular convex objects in E^4 are four-dimensional polyhedrons whose boundaries are built up by regular three-dimensional polyhedrons [Bö91]:

ε	κ	μ	ζ	
5	10	10	5	5-cell-body bounded by 5 tetrahedrons
16	32	24	8	8-cell-body bounded by 8 hexaedrons
8	24	32	16	16-cell-body bounded by 16 tetrahedrons
24	96	96	24	24-cell-body bounded by 24 octahedrons
600	1200	720	120	120-cell-body bounded by 120 dodecahedrons
120	720	1200	600	600-cell-body bounded by 600 tetrahedrons

2.1.5 Z^n as Incidence Structure

We can derive some further general conclusions from the matching formula. Assuming that these formulas for all values $k,l \leq n$ are valid, the formulas are also fulfilled for all $k,l \leq n-1$. Therefore, we get an $(n-1)$-dimensional incidence structure Σ_{n-1} from an n-dimensional incidence structure Σ_n by neglecting the n-dimensional elements in Σ_n. In that manner, the structure of figure 2.1.-5 is, by neglecting the cells, a two-dimensional incidence structure with $(\varepsilon,\varkappa,\mu,\nu,\lambda)=$ (16,32,24,4,4) which cannot be drawn in the plane without line crossing. Further, we get by neglecting both, cells and meshes, a regular (non-planar) graph with $\varepsilon=16$ points, $\varkappa=32$ edges, and with a neighborhood degree $\nu=4$. Therefore, we can formulate the following theorem:

Theorem 2.1.-1: Given an n-dimensional incidence structure Σ_n. Neglecting all k-dimensional elements with $n'<k\leq n$, we get an n'-dimensional incidence structure $\Sigma_{n'}$.

There is another possibility to derive a new incidence structure from the given one. If we choose a single n'-dimensional element of an n-dimensional structure, then once more a structure arises with $a_l'=b_{n'l}$ l-dimensional elements. The new structure constants are the same as the old because a k-dimensional element is attached to b_{kl} l-dimensional elements in both structures for $l<k$.

Theorem 2.1.-2: Given an n-dimensional incidence structure Σ_n. If we take a single n'-dimensional element with all of its attached k-dimensional elements for $k<n'$ then we obtain a new n'-dimensional incidence structure $\Sigma_{n'}$ with the same structure constants as in Σ_n. This incidence structure $\Sigma_{n'}$ contains b_{nl} l-dimensional elements with $l<n'$.

A third method to generate new incidence structures is given by a *duality principle*. If we change in figure 2.1.-1 the meshes and the points, we get once more incidence structures where $\nu'=\lambda$ and $\lambda'=\nu$. The structure with $(\nu,\lambda)=$ (4,4) is *self-dual*.

Theorem 2.1.-3: Given an n-dimensional incidence structure Σ_n. A new incidence structure Σ'_n is obtained when we replace the numbers of elements and the structure constants by

$$a_k' = a_{n-k} \qquad \text{and} \qquad b_{kl}' = b_{n-k,n-l} \ .$$

In all three cases described by theorems 2.1.-1/2/3, we can prove that the matching formula and the relationships between structure constants are fulfilled. In fact, therefore new incidence structures are generated.

There is a simple and very important case where n-dimensional finite and infinite incidence structures exist. The n-dimensional number lattice is such a structure. Assuming that $n=1$, we can close the structure in a ring-like manner so that a finite homogeneous structure of type Z^1 arises (see the second row in figure 2.1.-3). For $n=2$, we obtain the middle net of figure 2.1.-1 which can be closed toroidally for constructing a finite two-dimensional incidence structure of type Z^2. Finally, the three-dimensional number lattice Z^3 with integer-valued point coordinates is an infinite incidence structure with

$$\begin{array}{llll}
b_{30}=8 & b_{31}=12 & b_{32}=6 & b_{33}=1 \\
b_{20}=4 & b_{21}=4 & b_{22}=1 & b_{23}=2 \\
b_{10}=2 & b_{11}=1 & b_{12}=4 & b_{13}=4 \\
b_{00}=1 & b_{01}=6 & b_{02}=12 & b_{03}=8\ .
\end{array}$$

With respect to the previous theorems, we can restrict ourselves to the determination of the structure constants b_{0k} for $1 \le k \le n$. Any k-dimensional element attached to the origin as zero-dimensional element can be represented as an n-tupel of one's and zeros in which the 1 k-times occurs. There are $n!/(k!(n-k)!)$ different tupels of this kind when we only take positive coordinate values into account. If we consider that the 1's can have positive or negative signs, we obtain altogether

$$b_{0k}^{(n)} = 2^k \binom{n}{k} .$$

The number b_{kl} is given in Z^n for $k<l$ by all possibilities of $(n-k)$-tupels with $l-k$ signed 1's and $n-l$ 0's so that

$$b_{kl}^{(n)} = 2^{l-k} \binom{n-k}{l-k} = 2^{l-k} \binom{n-k}{n-l} .$$

The number b_{kl} with $k>l$ is the number of all l-dimensional elements attached to a k-dimensional element in Z^n. This number is independent of dimension n. Because of theorems 2.1.-2/3, we obtain

$$b_{kl}^{(n)} = b_{n-k,\,n-l}^{(n)} = 2^{k-l} \binom{k}{k-l} = 2^{k-l} \binom{k}{l} .$$

We condense these results to the *formula of structure constants* for Z^n:

<u>Formula 2.1.-2:</u>

$$b_{kl}^{(n)} = \begin{cases} 2^{l-k} \binom{n-k}{n-l} & \text{for } k<l \\ 1 & \text{for } k=l \\ 2^{k-l} \binom{k}{l} & \text{for } k>l \end{cases}$$

These expressions for the structure constants b_{kl} of the n-dimensional number lattice considered as incidence structure are fundamental for finite toroidally closed structures and for infinite grid point spaces both of type Z^n (see also [Ro71, Kl83]). We can show by very simple proofs that all matching formulas can be fulfilled using the structure constants of Z^n. Therefore all axioms, laws, theorems, and conclusions derived up to now for homogeneous incidence structures are valid for the Z^n, too. It shall be mentioned that with

Formula 2.1.-3:

$$\frac{b_{lk}}{b_{kl}} = \frac{2^{l-k}\binom{l}{k}}{2^{l-k}\binom{n-k}{n-l}} = \frac{\binom{n}{k}}{\binom{n}{l}} \qquad \text{for} \quad k \leq l \;,$$

$$\frac{b_{lk}}{b_{kl}} = \frac{2^{k-l}\binom{n-l}{n-k}}{2^{k-l}\binom{k}{l}} = \frac{\binom{n}{k}}{\binom{n}{l}} \qquad \text{for} \quad k \geq l$$

the important identity

Formula 2.1.-4:

$$\sum_{k=0}^{n} (-1)^k \frac{b_{lk}}{b_{kl}} = \frac{1}{\binom{n}{l}} \sum_{k=0}^{n} (-1)^k \binom{n}{k} = 0$$

is valid for the structure constants b_{kl} of Z^n for $n>0$ (this relationship will be used in section 2.4.2).

2.2 Oriented Neighborhood Structures

2.2.1 Orientation of a Neighborhood Structure

The theoretical/mathematical foundation of a scientific field requires two fundamental preconditions:

- the discipline (notions, methods, results) has to be well-developed,
- the mathematical tools necessary for formulating the theory have to be available.

Both preconditions are well satisfied for a topologically based theory of neighborhood structures as used in digital image processing.

At the level of a fundamental theory of image processing, some established branches of mathematics provide the tools, including set theory, graph theory, and theory of algebraic complexes. But the topological theory of neighborhood structures is motivated by typical problems of image processing, and the applied mathematical tools are always related to certain possible interpretations in the area of image processing.

N(p) is the set of all neighbors $q \in P$ of the point $p \in P$ in a neighborhood structure $[P,N]$, and $\nu_p = \nu(p) = card(N(p))$ is the number of all neighbors of p. By *cyclic ordering* of the ν_p neighbors, we obtain the *neighborhood cycle*

$$c(p) = \langle q_1, q_2, \ldots, q_{\nu_p} \rangle \quad .$$

An isolated point $p \in P$ gives the *empty cycle* $c(p) = \langle\,\rangle$. If p has only one neighbor q, it is $\nu_p = 1$, and the cycle is $c(p) = \langle q \rangle$.

<u>Def. 2.2.-1:</u> If we arrange cyclically all neighbors $q \in N(p)$ of each point $p \in P$ in a neighborhood structure $[P,N]$, an *oriented neighborhood structure* $[P,N,C]$ arises whereby C is the set of all cycles $c(p)$.

The procedure described in definition 2.2.-1 means that an *orientation* C is attached to the neighborhood structure $[P,N]$. For a given finite neighborhood structure, there exist

$$\prod_{p \in P} (\nu_p - 1)!$$

different orientations which may be specified on $[P,N]$.

Now it is possible to define edge sequences which are generated by such an orientation C. Any undirected edge $\{p,q\}$ of a neighborhood structure $[P,N]$ corresponds to two directed edges $(p,q),(q,p) \in N$.

Def. 2.2.-2: A directed edge $(q,r) \in N$ is the *successor* of the directed edge $(p,q) \in N$, and (p,q) is the *predecessor* of (q,r) if $c(q) = < \ldots p,r \ldots >$.

Thus, any directed edge (p,q) defines unambiguously a following edge and is defined unambiguously by a forerunning edge. Therefore (p,q) determines a sequence of directed edges which is finite and cyclic for finite neighborhood structures.

Def.2.2.-3: The finite cyclic edge sequence, defined by an edge $(p,q) \in N$ in an oriented neighborhood structure $[P,N,C]$, corresponds to an unambiguously defined cyclic path $m(p,q)$ through points of the neighborhood structure. This path is called a *mesh* or a *directed way*.

In figure 2.2.-1, an oriented neighborhood structure is shown whereby the cyclic ordering of neighbors $q \in N(p)$ is given by the clockwise ordering of edges outgoing from p. The edge (a,b) has the successor (b,e), the edge (b,e) has the successor (e,d) etc. so that finally the resulting mesh $m(a,b) = <a,b,e,d,c,h>$ arises.

Def. 2.2.-4: The number of points in a mesh m is called the length $\lambda(m)$ of the mesh. An isolated point is characterized by an *empty cycle*, and a mesh of length $\lambda = 0$ is attached to each isolated point.

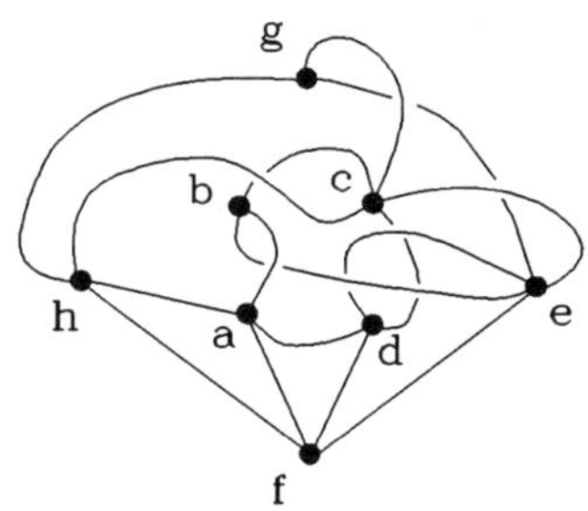

Fig. 2.2.-1: Oriented neighborhood structure

Because each edge (p,q) appears exactly once in any mesh, the sum of all mesh lengths is equal to the number of all directed edges. Therefore we obtain the following *mesh theorem* as the second important formula of oriented neighborhood structures apart from the vertex theorem 1.1.-1.

Theorem 2.2.-1: In an oriented neighborhood structure $[P,N,C]$, ω is the number of all different directed ways (meshes). Then

$$\sum_{m \in M} \lambda(m) = 2\varkappa \qquad \text{and} \qquad \bar{\lambda} = \frac{2\varkappa}{\omega}$$

are fulfilled with M as set of all meshes and $\bar{\lambda}$ as *mean way length* or *mean mesh length*.

2.2.2 Euler Characteristic of a Neighborhood Structure

For finite structures, we will derive an important quantitative relationship between the numbers of points, edges, and meshes. For a connected structure Σ_n with $n>0$, we have always

$$\varkappa \leq \frac{\varepsilon(\varepsilon-1)}{2}$$

because the corresponding neighborhood graph contains $\varepsilon(\varepsilon-1)/2$ edges in maximum. In the case of a two-dimensional incidence structure, the following theorem is fulfilled:

Theorem 2.2.-2: Let Σ_2 be a connected oriented neighborhood structure $[P,N,C]$ with ε points, $\varkappa$ edges, and ω meshes. Then

$$\varepsilon - \varkappa + \omega = 2(1-g) = \chi \leq 2$$

is fulfilled with g as *genus* of the structure and χ as *Euler characteristic* of the structure.

Structure	ε	$\varkappa$	ω	g	χ
	1	0	1	0	2
	2	1	1	0	2
	3	2	1	0	2
	3	3	2	0	2
	4	3	1	0	2
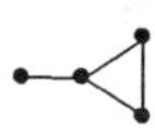	4	4	2	0	2
	4	5	3	0	2
	4	5	1	1	0
	4	6	2	1	0
	5	10	3	2	−2

Fig. 2.2.-2: Genus g and Euler characteristic χ of simple oriented neighborhood structures

For very simple structures, this theorem is fulfilled as we can see in figure 2.2.-2 for small numbers of ε. But it is also illustrated in figure 2.2.-2 that the more complicated the structures are the larger the genus and the smaller the Euler characteristic will become. Of course, all further examples and attempts to determine g and χ yield no proof for theorem 2.2.-2. We need exact mathematical conclusions based on definitions given up to now.

If we draw the examples of figure 2.2.-2, we work constructively by point-wise and edge-wise building-up a structure Σ in such a manner that the structure is always connected. Thereby, we use the following independent elementary steps:

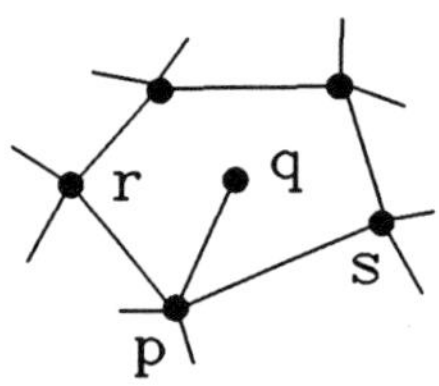

Fig. 2.2.-3: Case 1 of structure alteration (see text)

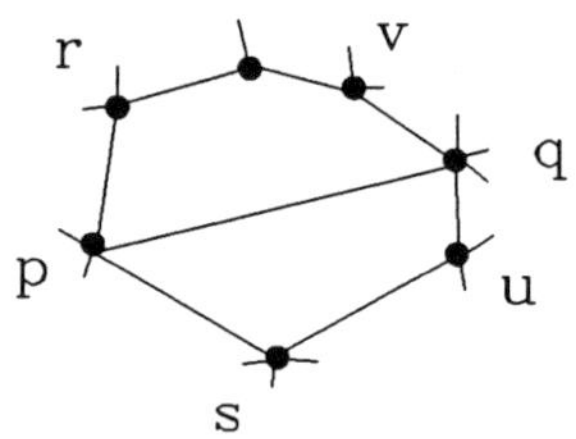

Fig. 2.2.-4: Example to case 2 of structure alteration

Case 1: We add a new point q to Σ and bind it to Σ by a new edge $\{q,p\}$ so that Σ' arises (figure 2.2.-3). Then we will get $\varepsilon'=\varepsilon+1$ and $\varkappa'=\varkappa+1$. Because q is a new neighbor of p, the neighborhood cycle $c'(p)= <\ldots r,q,s\ldots>$ arises from the old cycle $c(p)=<\ldots r,s\ldots>$. The neighborhood cycle of q is $c(q)=<p>$. The edge (r,p) is contained only in one mesh $m=<\ldots r,p,s,\ldots>$ of Σ. Because of that, we get the new mesh $m'=<\ldots r,p,q,p,s,\ldots>$, and this is the only changed mesh of Σ. Therefore it is $\omega'=\omega$ i.e. $g'=g$ and $\chi'=\chi$.

Case 2: A more complicated case is adding of a new edge $\{p,q\}$ between two already existing points of Σ in such a way that this edge lies within only one old mesh (figure 2.2.-4). For that, we have to insert p in the neighborhood cycle of q, and q in the neighborhood cycle of p. Now, we get new neighborhood cycles $c(p)=<\ldots r,q,s,\ldots>$ for point p and $c(q)=<\ldots u,p,v,\ldots>$ for point q. Therefore, from $m=<\ldots r,p,s,\ldots, u,q,v,\ldots>$ we obtain two new meshes, namely the mesh $m_1=<\ldots r,p,q,v,\ldots>$ and the mesh $m_2= <\ldots u,q,p,s,\ldots>$. For the new neighborhood structure Σ' we get $\varepsilon'=\varepsilon$, $\varkappa'=\varkappa+1$, and $\omega'=\omega+1$, i.e. $g'=g$ and $\chi'=\chi$.

Case 3: The last (third) elementary case is adding a new edge $\{p,q\}$ between two already existing points of Σ in such a way that this edge lies within two of the old meshes (figure 2.2.-5). For that, we have to integrate p into the neighborhood cycle of q, and q into the neighborhood cycle of p. We have then new cycles $c(p)=<\ldots r,q,s,\ldots>$ and $c(q)=<\ldots u,p,v,\ldots>$. Therefore from meshes $m_1=<\ldots r,p,s,\ldots>$, $m_2=<\ldots u,q,v,\ldots>$, we obtain the mesh $m=<\ldots r,p,q,v,\ldots u,q,p,s,\ldots>$. Now, Σ' is characterized by $\varepsilon'=\varepsilon$, $\varkappa'=\varkappa+1$, and $\omega'=\omega-1$, i.e. $g'=g+1$ and $\chi'=\chi-2$.

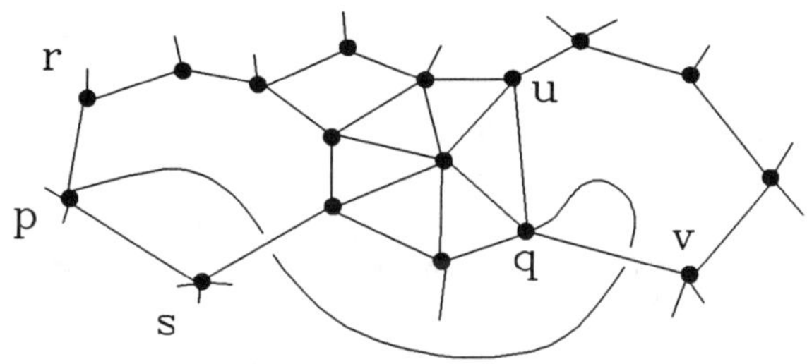

Fig. 2.2.-5: Example to case 3 of structure alteration (see text)

In table 2.2.-1, the elementary operations to construct and to pull down a two-dimensional neighborhood structure are shown. Using the operators F_1, F_2, F_3, and the corresponding inverse operators, we can handle all such structures by maintenance of connectivity.

Table 2.2.-1: Operators to construct two-dimensional structures

case	operator	Δg	$\Delta\chi$	operator	Δg	$\Delta\chi$
1	F_1	0	0	F_1^{-1}	0	0
2	F_2	0	0	F_2^{-1}	0	0
3	F_3	+1	-2	F_3^{-1}	-1	+2

In respect of this table, the Euler characteristic χ can be reduced only by steps of $\Delta\chi=2$ so that the theorem 2.2.-2 is fulfilled. From this constructive proof several conclusions may be derived. First, for the inverse process of deleting edges or points of a given structure we may conclude that for a resulting *substructure* the Euler characteristic cannot be smaller than the original one.

The operators F_i introduced here belong to the class of *Euler operators* used in computer aided modelling [Ba74,Br80, Ma82, Mä83, An85]. They are basic operators used for handling topological problems such as creating and deleting connected components (a face and a vertex, an edge and a face, an edge and a vertex etc.). In [Wu89], five three-dimensional construction operators and the five corresponding inverse operators are discussed.

Theorem 2.2.-3: If Σ' is a substructure of a finite connected structure Σ, then $\chi(\Sigma')\geq\chi(\Sigma)$. If the structure Σ consists of k connected components then $\chi(\Sigma)=\varepsilon-\varkappa+\omega\leq 2k$.

On the other hand, if $\chi(\Sigma)>2$ is given for a structure Σ then we can conclude that Σ is not connected but consists of several components not connected to one another.

By a re-orientation of any neighborhood cycle $c(p)$ of a structure Σ, a new structure Σ' arises in general. Such a re-orientation can be done by the operators of table 2.2.-1. First, we delete an edge $\{p,q\}$ so that $\Delta\chi\in\{0,+2\}$. Then we insert $\{p,q\}$ taking into account the new orientation. By this second step, we get $\Delta\chi\in\{-2,0\}$. Therefore the Euler characteristic is changed by an even number. Because ε and $\varkappa$ are unchanged, the variation of the mesh number $\Delta\omega$ caused by re-orientation is changed by an even number, too.

Theorem 2.2.-4: A given neighborhood structure Σ belongs independently of its orientation either in the structure class characterized by an even mesh number ω or in the class characterized by an odd mesh number ω.

Because the singular character of $\chi=2$ for connected structures, one can ask whereby such structures are characterized otherwise. Now, we can draw in such structures always an edge between two points within a mesh, i.e. without crossing other edge lines. Such structures are named *planar*. Planar structures can be drawn without edge crossing in the Euclidian plane and at the surface of a sphere, respectively.

Def. 2.2.-5: A two-dimensional connected incidence structure Σ_2 with $\chi(\Sigma)=2$ is denoted as *planar structure*. The corresponding neighborhood graph Σ_1 as skeleton of Σ_2 is then a *planar graph*. A graph is then planar if there is an orientation so that the resulting oriented neighborhood structure is planar.

In 1930, Kuratowski proved the following theorem [Ku30, Sa72]:

Theorem 2.2.-5: A graph is *nonplanar* if it contains a subgraph which is G_5 or $G_{3,3}$ shown in figure 2.2.-6.

With respect to definition 2.2.-5 and assuming that G_5 is planar, $\omega=2+\varkappa-\varepsilon=7$ follows for G_5 . According to theorem 2.2.-1, then we obtain a mean mesh

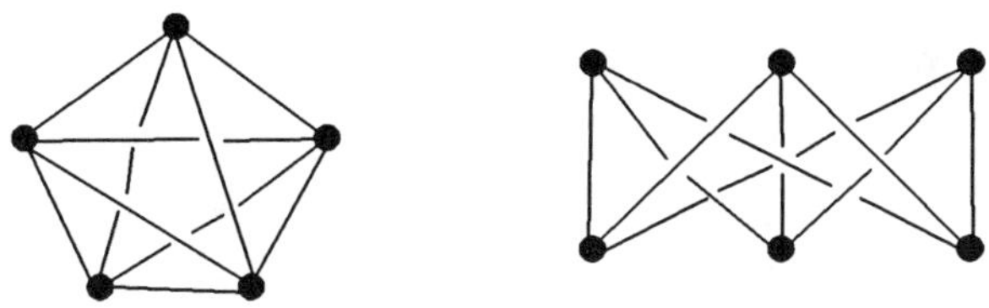

Fig. 2.2.-6: Kuratowki's nonplanar graphs G_5 and $G_{3,3}$

length $\bar{\lambda} = 2\varkappa/\omega = 20/7$ which is impossible because any mesh has at least the length 3. Therefore G_5 is nonplanar. Likewise for $G_{3,3}$, we obtain $\omega = 2 + \varkappa - \varepsilon = 5$ assuming that $G_{3,3}$ is planar. Because any mesh on $G_{3,3}$ must have the length 4 at least, the resulting mean value $\bar{\lambda} = 2\varkappa/\omega = 18/5$ is in contradiction to that, and therefore the graph $G_{3,3}$ is nonplanar, too.

2.2.3 Border Meshes and Separation Theorem

It is a central idea of image processing to separate image segments for further analysis. This separation may be realized by characterizing boundary meshes (or border meshes) of substructures which describe image segments. Following this usual approach, the study of boundary meshes may be performed at the abstraction level of oriented neighborhood structures.

Let M be a subset of the point set P of the structure $\Sigma = [P,N,C]$ with $\overline{M}$ as its complement $P-M$. Such a set M determines a division of the set N of edges into three subsets

- the set N_M of inner edges (p,q) with $p,q \in M$,
- the set $N_{\overline{M}}$ of outer edges (p,q) with $p,q \in \overline{M}$,
- and the set N_C of edges (p,q) with $p \in M$ and $q \in \overline{M}$ of edges wich connect the both sets M and $\overline{M}$ (so-called border edges or *boundary* edges).

Now, let m be a mesh of the substructure Σ_M generated from structure Σ by set M (figure 2.2.-7). Such a mesh is either a *core mesh* of Σ_M (if m is also a mesh of structure Σ) or a *border mesh* of Σ_M (if m cannot be found in the structure Σ). If (p,q) and (q,r) are two successive edges of a boundary mesh of Σ', and if we have the cycle $c(q) = \langle \ldots p \ldots s \ldots q \ldots \rangle$ for q, then we denote (q,s) with $q \in M$ and $s \in P-M$ as an *attached edge* of mesh m. We will denote the

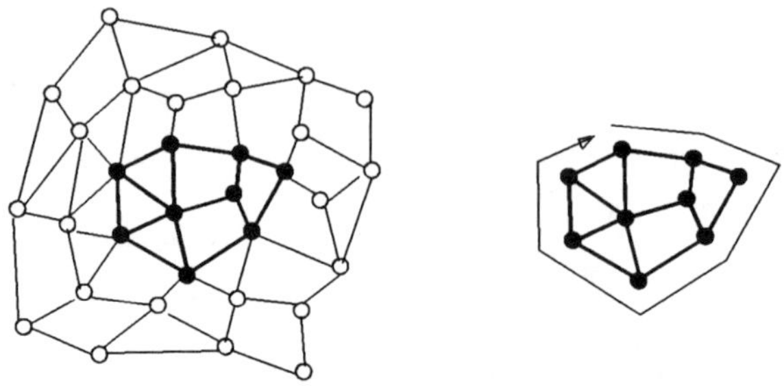

Fig. 2.2.-7: Core meshes and border meshes of substructures

number of core meshes by μ, and the number of border meshes by β. Therefore, it follows for the number of all meshes or for the number of all directed ways the relationship $\omega=\mu+\beta$.

Given a boundary mesh m of substructure Σ_M, we can delete all edges attached to m. Then, corresponding to operators $\boldsymbol{F}_2^{-1}$ and $\boldsymbol{F}_3^{-1}$, the Euler characteristic of the structure will be changed by $\Delta\chi\geq 0$. Finally, we have deleted all attached edges of m but one (figure 2.2.-8). In this moment, there exists a mesh m' generated by edge (q,p) as $m'=<\ldots u,q,p,s\ldots r,p,q,v\ldots>$.

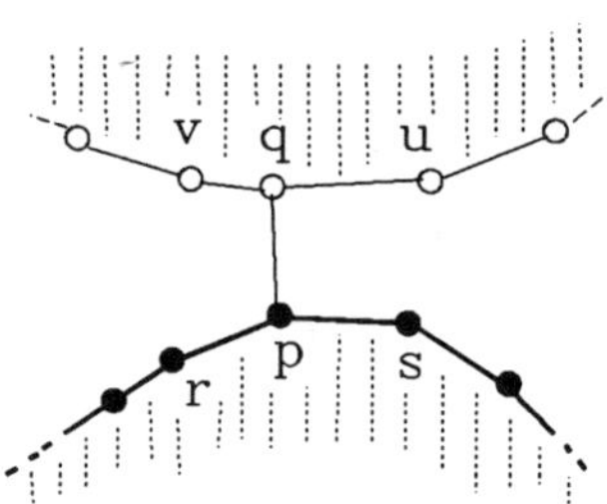

Fig. 2.2.-8: End configuration for proving the border mesh theorem

After deletion of the last edge (p,q) attached to boundary mesh m, the mesh m' is split into the boundary mesh $m=<\ldots r,p,s\ldots>$ and yet another mesh $m''=<\ldots u,q,v\ldots>$. Therefore the number $\varkappa$ of edges is decreased by 1, and the number ω of meshes is increased by 1. Thus, the Euler characteristic of the new generated structure is actually greater than before, i.e. $\chi(\Sigma')>\chi(\Sigma)$, and we get the following *border mesh theorem*:

<u>Theorem 2.2.-6:</u> Let Σ_M be a proper sub-structure of a finite structure $\Sigma=[P,N,C]$ with $M \subset P$, and let the structure Σ' be constructed out of Σ by deleting of all boundary edges attached to a certain border mesh of Σ_M. Then it is always $\chi(\Sigma') > \chi(\Sigma)$.

Especially for connected structures with $\chi=2$, this theorem leads immediately to the following discrete analog of Jordan's curve theorem:

<u>Theorem 2.2.-7:</u> Let Σ' be constructed out of a finite connected structure Σ by deleting of all boundary edges attached to a certain border mesh of Σ. Assuming that $\chi(\Sigma)=2$, i.e. Σ is a so-called *planar structure*, $\chi(\Sigma')>2$ follows, and Σ' consists of at least two components.

This *separation theorem* ensures that any border mesh of a substructure Σ_M of a planar connected neighborhood structure $\Sigma=[P,N,Z]$ separates Σ_M and at least one of its complementary components. Because this separation theorem is not provable in case of non-planar structures (as for example for the so-called 8-neighborhood), the use of non-planar support structures should be avoided in image processing. Especially for the triangular structure $(\nu,\lambda)=(6,3)$ of figure 2.1.-1 it may be stated that any border mesh of a substructure separates exactly one complementary component [Vo88].

Because the border meshes are of particular importance for image processing, we will give here an algorithm to determine border meshes in oriented neighborhood structures. This algorithm is a generalization of the common algorithm for contur following [Vo88].

<u>Alg. 2.2.-1:</u> Determination of border meshes

1. Be given the oriented neighborhood structure $\Sigma=[P,N,C]$ and a subset $M \subset P$. We assume $card(M)>1$.
2. Get a list L containing all border pairs (p,q) with $p \in M$ and $q \notin M$.

    3. If the list L is non-empty then go to step 4. Otherwise all border meshes are found.
       "*** end of algorithm ***"
    4. Choose a border pair $(p,q) \in L$ and delete it in L. Set $(p,q) \rightarrow (p_0,q_0)$. The point p_0 is the first point of a border mesh.

        5. Determine the neighbor $r \in N(p)$ which follows $q \in N(p)$ in the neighborhood cycle $c(p)= <\ldots q,r,\ldots>$. If $r \in M$ then go to step 7.

6. Otherwise, the border pair (p,r) gets the same border mesh as (p,q). Delete (p,r) in L, set $r \to q$, go to step 5.
7. The pair (p,r) is an edge of the border mesh determined by (p_0,q_0), and $r \in M$ is a further point of this border mesh. Determine the neighbor $s \in N(r)$ which follows $p \in N(r)$ in the neighborhood cycle $c(r) = <\dots p,s,\dots>$. If $s \in M$ then set $(r,s) \to (p,r)$, and go to step 7.
8. Otherwise, the border pair (r,s) is a further border pair of the border mesh determined by (p_0,q_0). Delete (r,s) in L, and set $(r,s) \to (p,q)$. If $(p,q) \neq (p_0,q_0)$ then go to step 5.

9. Otherwise, the border mesh determined by the border pair (p_0,q_0) is determined. Go to step 3.

2.2.4 Search in Oriented Neighborhood Structures

Now, we will here generalize the graph search problem (search in one-dimensional incidence structures) as investigated in section 1.3.2. There was a list L given to describe the neighborhood graph by $L=((a,N(a)), (b,N(B)),\dots)$ where the sublists contain for each point p of the graph the corresponding set $N(p)$ of neighbored points. The order of neighbors in $N(p)$ was meaningless.

In oriented neighborhood structures, we have to take into account that the list L is given by $L=((a,c(a)), (b,c(b)),\dots)$ where $c(p)$ is the neighborhood cycle of point p. The problem is to determine the topological structure of finite connected oriented neighborhood structures $\Sigma=[P,N,C]$, that means the determination of the number ε of points, the number $\varkappa$ of edges, and the number ω of directed ways (meshes). From these quantities, we obtain the Euler characteristic $\chi=\varepsilon-\varkappa+\omega$ of the structure Σ. This task is up to now not yet investigated explicitly in image processing. Only in [Do87], the traversing of the cells of a cellular complex is discussed.

However the work in surface detection of three-dimensional objects goes in that direction (see chapter 3.1). Thereby, it is not the main problem to find any directed way because with a directed starting edge (p,q) the next following edge (q,r) is determined by the neighborhood cycle $c(q)=<p,r,s,\dots>$ etc. Thus, an edge sequence is found up to the starting point p is reached again.

The problem is that the directed way $<p,q,r,s,\dots>$ determined by the edge (p,q) is determined also by the other edges $(q,r),(r,s),\dots$ Therefore, we have to use lists for storing the meshes already found. The general algorithm 2.2.-2 describes this approach.

Alg. 2.2.-2: Structure search by means of lists

1. Given the list $L=((a,c(a)),(b,c(b)),\ldots)$ of structure $\Sigma=[P,N,C]$ with all points $p\epsilon P$ and all neighborhood cycles $c(p)\epsilon C$.
2. Build up a list $K=((p_1,q_1),(p_2,q_2),\ldots)$ of all directed edges $(p_i,q_i)\epsilon N$.
   3. If K is empty then go to step 7.
   4. Select any edge (p,q) out of list K.
   5. Determine the directed way $<p,q,r,\ldots>$, and delete all directed edges $(p,q),(q,r),\ldots$ of this way in list K.
   6. Go to step 3.
7. All directed ways of Σ are found.

We have already mentioned in section 1.3.2 that this principal method is ineffective because of the cumbersome search in lists. In image processing, we can address points, neighbors, edges etc. by their coordinate values so that the access to these data is more time effective. Certainly, then the neccessity arises to "mark" the points etc. in any manner. For edges in Z^3, this problem is discussed in section 3.1.3.

Alg. 2.2.-3: Structure search by means of marking

```
"A structure Σ=[P,N,C] is given where the data can be addressed directly.
Let p∈P be the start point"
0 → ϰ
0 → ω
INIT_STACKP                    "for points"
PUSHP p                        "point p in point stack"
1 → ε
while not EMPTY_STACKP do
begin
   POPP p                      "select point p from point stack"
   INIT_STACKD                 "for directed edges (p,q)"
   for all neighbors q of p do
   begin
        if point q is not marked then
        begin
             PUSHP q
             ε+1 → ε
             MARKP q           "mark point q"
        end
        if directed edge (p,q) is not marked then
        begin
             MARKD (p,q)       "mark directed edge"
             ϰ+1 → ϰ
        end
        p → r
```

```
            q → s
            while not s=p do
            begin
                NEXT_EDGE (r,s) (s,t)
                if directed edge (s,t) is not marked
                then
                begin
                    MARKD (s,t)
                    ϰ+1 → ϰ
                end
                s → r
                t → s
            end  "of directed way"
            ω+1 → ω
        end  "of treatment of point p"
    end  "of structure"
    ϰ/2 → ϰ                          "number of undirected edges"
    ε-ϰ+ω → χ                        "Euler characteristic"
    "end of algorithm"
```

The used procedures **INIT_STACKP, PUSHP, EMPTY_STACKP, POPP, MARKP, INIT_STACKD,** and **MARKD** have the same meaning as in section 1.3.3 discussed. The new procedure **NEXT_EDGE** determines the edge (s,t) following edge (r,s) in respect to the neighborhood cycle $c(s)=<r,t,\ldots>$. The marking of points of the neighborhood structure - they correspond to the marginal edges of Z^3 - can be done by the edge marking discussed in section 3.1.3. But because we have to mark here also directed edges of the neighborhood structure - they correspond to marginal meshes in Z^3 - one has to develop appropriate algorithms for that task.

2.2.5 Coloring in Oriented Neigborhood Structures

As in section 1.3.5, we will investigate regular coloring in oriented neighborhood structures. We denote the structure $\Sigma=[P,N,C]$ as *regularly colorable* with c colors if any two neighbored points - i.e. two points which incides with the same edge - have different colors. The smallest number $c(\Sigma)$ of colors which is necessary for a regular coloring of Σ is denoted as *chromatic number* of Σ.

If the neighborhood graph $\Gamma=[P,N]$ corresponding to the oriented neighborhood structure $\Sigma=[P,N,C]$ is a complete graph with ε points (see section 1.1.3), we need also $c(\Sigma)$ colors for a regular coloring.

The Euler characteristic $\chi(\Sigma)=\varepsilon-\varkappa+\omega$ of a structure with complete neighborhood graph cannot be greater than a defined limit [Vo88]:

<u>Theorem 2.2.-8:</u> Let $\Sigma=[P,N,C]$ be a structure with complete neighborhood graph $\Gamma=[P,N]$ and $\varepsilon=card(P)\geq 3$. If we assume that $\lambda(m)\geq 3$ for the length λ of all meshes m, then $\chi(\Sigma)\leq -\varepsilon(\varepsilon-7)/6$ follows.

This theorem can be proved using $\varkappa=\varepsilon(\varepsilon-1)/2$ for a complete graph and using the mesh theorem as

$$3\omega \leq \bar{\lambda}\omega = \sum_m \lambda(m) = 2\varkappa = \varepsilon(\varepsilon-1) .$$

From $\omega\leq\omega_{max}=\varepsilon(\varepsilon-1)/3$, we obtain

$$\chi(\Sigma) \leq \varepsilon - \varkappa + \omega_{max}$$

$$= \varepsilon - \frac{\varepsilon(\varepsilon-1)}{2} + \frac{\varepsilon(\varepsilon-1)}{3} = -\frac{\varepsilon(\varepsilon-7)}{6} .$$

From this inequality, it follows for $\varepsilon>3$

$$\varepsilon \leq \varepsilon_{max}(\chi) = ent\left[\frac{7+\sqrt{49-24\chi}}{2}\right] ,$$

i.e. the point number ε of a neighborhood structure Σ with a complete neighborhood graph and given $\chi(\Sigma)$ cannot be greater than ε_{max}.

Now, let be given structures Σ with Euler characteristic χ. The smallest of all these structures - smallest in respect of the point number ε - consists of ε_{max} points and possesses a complete neighborhood graph. Therefore we can conclude that at least $c_{min}=\varepsilon_{max}$ colors are needed for a regular coloring of the structure (theorem 2.2.-9, table 2.2.-2).

<u>Theorem 2.2.-9:</u> The necessary number c_{min} of colors for regular coloring of a connected structure $\Sigma=[P,N,C]$ with an Euler characteristic $\chi=\chi(\Sigma)$ is given by

$$c_{min}(\chi(\Sigma)) = ent\left[\frac{7+\sqrt{49-24\chi}}{2}\right] .$$

A sufficient number c_{max} of colors for regular coloring of connected structures can be found if we estimate the mean neighborhood degree of the corresponding neighborhood graph $\Gamma=[P,N]$. As we have seen, it is $\omega\leq 2\varkappa/3$. Therefore, it follows

$$\chi = \varepsilon - \varkappa + \omega \leq \varepsilon - \frac{\varkappa}{3} .$$

This inequality yields $\varkappa \leq 3(\varepsilon - \psi)$ or

$$\overline{\nu} = \frac{2\varkappa}{\varepsilon} \leq \frac{6(\varepsilon - \chi)}{\varepsilon} = 6 - \frac{6\chi}{\varepsilon} .$$

Table 2.2.-2: Color number c_{min} for given Euler characteristic χ

χ	c_{min}	χ	c_{min}
2	4	-12	12
0	7	-14	13
-2	8	-16	13
-4	9	-18	14
-6	10	-20	15
-8	11	-22	15
-10	12	-24	16

If the mean value of any integer numbers is not greater than a real number r, then at least one of the integers has to be smaller than or equal to r. The neighborhood degrees ν are integers. It follows that there is at least one point $p \in P$ in a structure $\Sigma = [P, N, C]$ with given Euler characteristic χ and $\varepsilon = card(P)$ for which the neighborhood degree $\nu(p)$ is not greater than the mean neighborhood degree. We denote $\nu_s = ent(6 - 6\chi/\varepsilon)$ as *specific neighborhood degree* of Σ.

For structures with given Euler characteristic χ, the specific neighborhood degree is dependent on ε. It shall be $\nu_s(\chi)$ the greatest specific neighborhood degree with respect to all possible values of ε.

It is $\nu_s(2) < 6$ because this value arises if the point number ε arises. Further, we obtain $\nu_s(0) = 6$. For structures with Euler characteristic $\chi < 0$, the value $\nu_s(\chi, \varepsilon)$ arises if ε decreases. In these cases, the greatest values will be yielded for the minimal possible point numbers ε.

For given ε and χ, the largest mean neighborhood degree follows for the edge number $\varkappa = \varepsilon(\varepsilon - 1)/2$, i.e. we obtain the value $2\varkappa/\varepsilon_{max} = \varepsilon_{max} - 1$ with ε_{max} as the greatest point number possible for complete structures with given χ:

$$\overline{\nu} \leq \varepsilon_{max} - 1 \leq \nu_{max}(\chi) = ent\left[\frac{5 + \sqrt{49 - 24\chi}}{2}\right] .$$

Because of this inequality, we obtain the following theorem.

Theorem 2.2.-10: The mean neighborhood degree of any structure with Euler characteristic $\chi<0$ is not greater than $\nu_{max}(\chi)$. It is $\nu_{max}(2)=5$ and $\nu_{max}(0)=6$. For $\chi'<\chi$ follows $\nu_{max}(\chi')\leq\nu_{max}(\chi)$.

Now, we can derive an algorithm for regular coloring of structures with Euler characteristic χ which needs only $\nu_{max}(\chi)+1$ colors.

Alg. 2.2.-4: Regular coloring for neighborhood graph Γ of structure Σ

1. Take $0\to k$, $\Gamma\to\Gamma'$, $\Sigma\to\Sigma'$, and $\nu_0=\nu_{max}(\chi(\Sigma))$.
2. There is at least one point p in Γ' with $\nu(p)\leq\nu_0$.
3. Take $k+1\to k$ and $p\to p_k$.
4. Delete p and all edges $\{p,q\}$ in Σ'. The arising structure Σ'' has an Euler characteristic $\chi(\Sigma'')\geq\chi(\Sigma')$ because we have used the inverse operators of table 2.2.-1.
5. Take $\Sigma''\to\Sigma'$ and $\Gamma''\to\Gamma'$. If there are yet points in Γ', go to step 2 of the algorithm.
6. The last deleted point shall be p_K. Now we construct the original structure Σ adding the points p_K, p_{K-1}, p_{K-2},... in reverse order. Each new point p_i have to be connected with at most ν_0 already added points. Therefore, we can color p_i with one of ν_0+1 colors.

Theorem 2.2.-11: Each structure Σ can be regularly colored with at most

$$c_{max}(\chi(\Sigma)) = ent\left[\frac{7+\sqrt{49-24\chi}}{2}\right]$$

different colors.

We see that $c_{min}=c_{max}$ for all structures with $\chi\leq0$. This result was conjectured by Heawood already in 1890 [He90]. But not before 1968, the theorem 2.2.-11 could be proved by Ringel and Youngs [Ri68, Sa72].

Therefore, the problem of regular coloring is solved for all structures with Euler characteristic $\chi\leq0$. For $\chi=2$, we have the inequalities $c_{min}=4\leq c\leq c_{max}=6$ (see theorem 1.3.-2). Appel and Haken have proved by means of a computer that also for $\chi=2$ the relationship $c_{max}=c_{min}=4$ is valid (see section 1.3.5). But the search for a simple proof of $c(2)=4$ goes on.

2.3 Homogeneous Oriented Neighborhood Structures

2.3.1 Homogeneity in Neighborhood Structures

A two-dimensional oriented neighborhood structure $\Sigma=[P,N,C]$ is *homogeneous* if for any point $p \epsilon P$ we have $\nu(p)=\nu=\text{const}$, and for any mesh m of the structure $\lambda(m)=\lambda=\text{const}$ is fulfilled. In figure 2.1.-1, the three possible structures $\Gamma(\nu,\lambda)$ are shown. These three structures and other homogeneous structures can be derived from the three topological main formulas which are given by the vertex theorem 1.1.-1, the mesh theorem 2.2.-1, and the Euler theorem 2.2.-2:

$$\varepsilon\,\nu = 2\,\varkappa\ ,$$

$$\omega\,\lambda = 2\,\varkappa\ ,$$

$$\varepsilon - \varkappa + \omega = \chi\ .$$

Now, we have three equations for five unknowns and given Euler characteristic χ. But because all unknowns must be positive integer numbers, a Diophantin equation system must be solved. Using $\varepsilon/\varkappa=2/\nu$ and $\omega/\varkappa=2/\lambda$, we can replace two unknowns and get the equation

$$\frac{2}{\nu} + \frac{2}{\lambda} = 1 + \frac{\chi}{\varkappa} \quad \rightarrow \quad \varkappa = \frac{\chi}{\dfrac{2\,(\nu+\lambda)}{\nu\,\lambda} - 1}\ .$$

Because the number $\varkappa$ of edges has to be finite, positive, and greater than 1, we obtain for $\chi=2$ the inequality

$$0 < \frac{2\,(\nu + \lambda)}{\nu\,\lambda} - 1 < 2 \quad \rightarrow \quad \frac{1}{2} < \frac{1}{\nu} + \frac{1}{\lambda} < \frac{3}{2}\ .$$

By a systematic examination of all possible solutions $\varkappa=\varkappa(\nu,\lambda)$, we get the values shown in table 2.3.-1.

The solutions $(\nu,\lambda,\varkappa,\varepsilon,\omega)=(c,2,c,2,c)$ consist of two points connected with c edges. Such structures are not neighborhood structures because there are multiple edges. They are unsuitable for image processing.

Another set of possible solutions is given by $(\nu,\lambda,\varkappa,\varepsilon,\omega)=(2,c,c,c,2)$ with any integer values c. They are are ring-like structures which are the support structures for one-dimensional image processing (see section 5.1.1).

As finite two-dimensional planar neighborhood structures, only the five structures of table 2.3.-2 and figure 2.3.-1 are meaningful for us.

Table 2.3.-1: Integer values $\varkappa=\varkappa(\nu,\lambda)$ for two-dimensional planar structures

	λ=1	2	3	4	5	6	7	8
ν=1								
2		2	3	4	5	6	7	
3		3	6	12	30			
4		4	12					
5		5	30					
6		6						
7		7						
8		8						

Table 2.3.-2: Two-dimensional finite planar neighborhood structures $(\nu,\lambda,\varkappa,\varepsilon,\omega)$

ν	λ	ε	$\varkappa$	ω	structure type
3	3	4	6	4	tetrahedron
3	4	8	12	6	hexahedron
3	5	20	30	12	dodecahedron
4	3	6	12	8	octahedron
5	3	12	30	20	icosahedron

Regarding all solutions of the Diophantin equation system for $\chi=2$, we have obtained the result that there are no planar homogeneous two-dimensional neighborhood structures suitable for image processing because we cannot say that images with maximally 20 points are of any practical interest.

But for infinite values of $\varepsilon,\varkappa,\omega$ there are the homogeneous solutions (ν,λ) of figure 2.1.-1. The question arises whether there are large finite neighborhood structures which are nonplanar, i.e. where we have $\chi\neq2$. Now, also for $\chi<0$, we find only structures with a small number of points (see table 2.3.-3 for $\chi=-2$, a table for $\chi=-4$ is given in [Vo88]).

2.3.2 Toroidal Nets

Now, we have the dilemma that the obtained structures are either homogeneous and planar (with validity of separation theorem) but with small point numbers, or they are non-homogeneous and therefore unsuitable for detailed theoretical investigations with respect to practical applications.

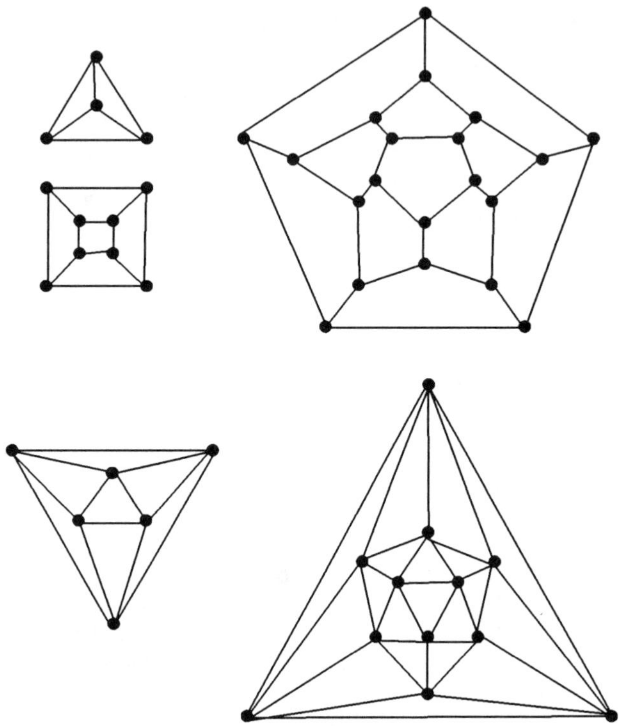

Fig. 2.3.-1: The five homogeneous planar neighborhood structures

Table 2.3.-3: Two-dimensional finite neighborhood structures $(\nu, \lambda, \varkappa, \varepsilon, \omega)$ for $\chi = -2$

ν	λ	ε	$\varkappa$	ω
3	7	28	42	12
	8	16	24	6
	9	12	18	4
	10	10	15	3
	12	8	12	2
	18	6	9	1
4	5	10	20	8
	6	6	12	4
5	4	8	20	10
7	3	12	42	28

There are two ways to overcome these difficulties. The first is to use the infinite planar structures $\Gamma(\nu,\lambda)$ of figure 2.1.-1 whereby the loss of homogeneity is shifted to infinity. That means to neglect effects caused by the limiting borders of finite images. The second way is to guarantee an overall homogeneity and to assume planarity only for (large enough) parts of finite structures.

The first way is used widely in practical image processing, and the second way in two-dimensional signal processing where periodicities of images in both directions is assumed for Fourier transforms.

The only possibility to find large finite homogeneous structures is given by the assumption $\chi=0$. In this case, the characteristic equation $2/\nu+2/\lambda=1+\chi/\varkappa$ is

Formula 2.3.-1:

$$\frac{2}{\nu} + \frac{2}{\lambda} = 1 \ .$$

This equation has the only integer valued solutions $(\nu,\lambda)=(6,3)$, $(4,4)$, and $(3,6)$. The corresponding structures are nonplanar. But they can be drawn at the surface of a torus without edge crossing, so that we speak of *toroidal structures* or *toroidal nets*. Toroidal structures do not lead to restrictions in size because with $\varepsilon-\varkappa+\omega=0$ also $c\varepsilon-c\varkappa+c\omega=0$ is fulfilled, and the structure chosen can be as large as we want (see table 2.3.-4).

Table 2.3.-4: Two-dimensional toroidal structures

ν	λ	ε	$\varkappa$	ω	remarks	
3	6	$2c$	$3c$	c	hexagonal nets,	$c\geq 3$
4	4	c	$2c$	c	orthogonal nets,	$c\geq 5$
6	3	c	$3c$	$2c$	triagonal nets,	$c\geq 7$

The most simple toroidal nets can be obtained by the minimal c's given in table 2.3.-4 (see [Vo88]). Therefore, we have found homogeneous oriented neighborhood structures suitable for two-dimensional image processing. Because of theoretical reasons, finite toroidal incidence structures are also in the n-dimensional case the optional way to introduce large homogeneous structures and to avoid difficulties with infinity.

2.3.3 Curvature of Border Meshes in Toroidal Nets

Now we will investigate objects in the three two-dimensional homogeneous neighborhood structures of figure 2.1.-1. For that, we use some definitions and results of sections 2.4.1/2 given for general n-dimensional objects.

Let $\Gamma=\Gamma(\nu,\lambda)$ be a finite toroidal structure with $(\nu,\lambda)=(3,6)$, $(4,4)$, $(6,3)$ as shown in figure 2.1.-1. Further, B shall be a connected set of points (i.e. an "object") with only one border mesh ($\omega=\mu+1$). The length of this border mesh shall be l.

We delete all k boundary pairs as mentioned in the border mesh theorem 2.2.-6. The remaining structure Σ' has an Euler characteristic $\chi(\Sigma')>\chi(\Sigma)=0$. Otherwise, it is $\chi(\Sigma')\leq 2$. Therefore, it follows

$$0 \leq \chi_B = \chi(\Sigma') \leq 2 \quad .$$

The three topological basic equations are

$$\text{eq1:} \qquad \nu\varepsilon - k = 2\varkappa \quad ,$$

$$\text{eq2:} \qquad \lambda\mu + l = 2\varkappa \quad ,$$

$$\text{eq3:} \qquad \varepsilon - \varkappa + \mu + 1 = \chi_B \quad .$$

The first equation follows because each of the ε object points has exactly ν neighbors but k of the edges connecting the object points with neighbors were removed as boundary pairs. The second equation follows because each of the μ core meshes of B has the length λ but the only border mesh has the length l. These two equations are special cases of the general object matching formula 2.4.-1.

A simple computation yields from the expression $\nu\lambda\cdot\text{eq3}-\lambda\cdot\text{eq1}-\nu\cdot\text{eq2}$ the equation

$$\lambda k - \nu l = \nu\lambda(\chi_B - 1) - (2\lambda + 2\nu - \nu\lambda)\varkappa \quad .$$

The factor before $\varkappa$ is identical to zero for toroidal structures $\Gamma(\nu,\lambda)$ because

$$2\lambda + 2\nu - \nu\lambda = 2\nu\lambda\left(\frac{1}{\nu} + \frac{1}{\lambda} - \frac{1}{2}\right) = 0 \quad .$$

Therefore, it follows

$$\frac{k}{\nu} - \frac{l}{\lambda} = \chi_B - 1 \quad .$$

Because χ_B is 0 or 2, we get for the expression $k/\nu - l/\lambda$ only two possible values namely $+1$ or -1.

Theorem 2.3.-1: In finite toroidal neighborhood structures $\Gamma(\nu,\lambda)$, the relationship

$$t = \frac{k}{\nu} - \frac{l}{\lambda} \in \{-1, +1\}$$

is fulfilled for each border mesh whereby for object conturs (outer border meshes) $t=+1$, and for inner border meshes (conturs of holes) $t=-1$.

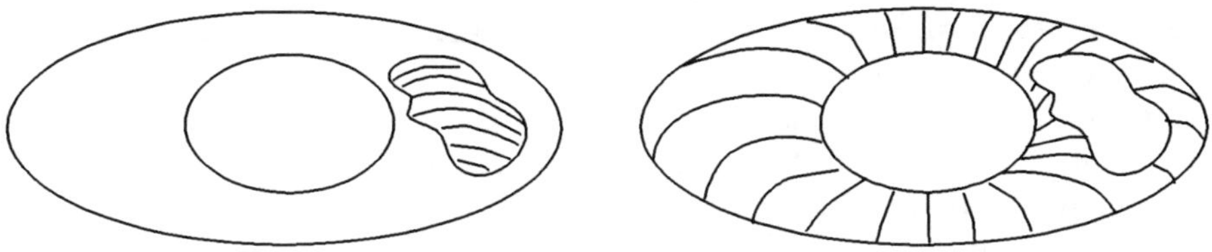

Fig. 2.3.-2: Planar object (left) with outer border mesh and toroidal object (right) with inner border mesh

If we have a planar object, the single border mesh corresponds to the object contur. But if the object is characterized by $\chi_B=0$ (i.e. B is a toroidal object), the border mesh includes planar regions and is therefore an inner boundary. Because all border meshes of an object are independent from one another, theorem 2.3.-1 is valid for each border mesh. We denote the quantity $t=k/\nu - l/\lambda$ as *total curvature* of a border mesh (see [Vo88]).

The fact that we can differentiate between inner and outer border meshes or between "in" and "out" in respect to an object is only given in toroidal nets $\Gamma(\nu,\lambda)$. A small land in a worldwide ocean on a spherical surface is an island, and the ocean is "out" of the island, the island lies "in" the ocean. Enlargement of the island - more and more - will transform the island to a worldwide continent, and the ocean will become a small sea. Now, the water is "in" of the continent. From a topological standpoint, a differentiation between "inside" and "outside" on a spherical surface is not possible.

Using the matching formulas eq1 and eq2, we obtain

$$\nu\varepsilon - k = \lambda\mu + l \quad .$$

Otherwise, it follows for an object with a single outer border mesh using $k=\nu(1+l/\lambda)$ the relationship

Formula 2.3.-2:

$$\varepsilon = \frac{\lambda}{\nu}\mu + l\left(\frac{1}{\nu} + \frac{1}{\lambda}\right) + 1 = \frac{\lambda}{\nu}\mu + \frac{l}{2} + 1 \quad .$$

This equation shall be denoted as *Pick's formula* [Vo88] because Pick has investigated grid point polygons of the two-dimensional orthogonal lattice already in 1899 [Pi99]: The area A of a proper grid point polygon (the corner points of the polygon have integer coordinate values) is determined by $A=i+b/2-1$. Here, i is the number of grid points inside the polygon, and b is the number of grid points at the boundary of the polygon. In figure 2.3.-3, we count $i=8$ and $b=7$ so that $A=21/2$ follows.

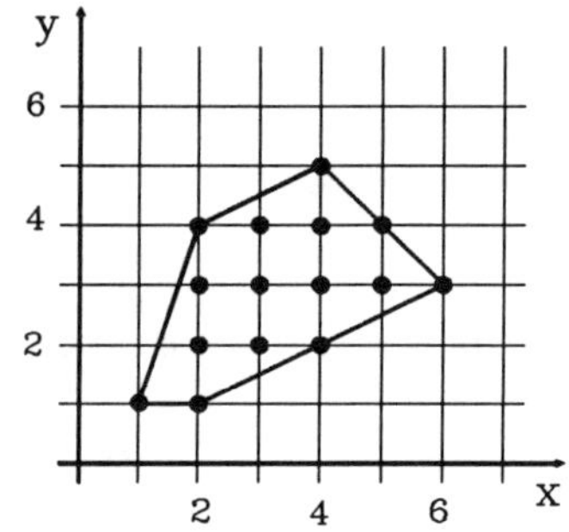

Fig. 2.3.-3: Example for Pick's formula

For a two-dimensional object with ß border meshes (i.e. one outer border mesh and ß-1 inner border meshes), the Euler characteristic is given by $\chi=\varepsilon-\varkappa+\mu+\beta=2$ with μ as number of core meshes. It follows then for the Euler number ψ of this object

$$\psi(B) = \varepsilon - \varkappa + \mu = 2 - \beta = 1 - H$$

where $H=\beta-1$ is the number of inner border meshes of the object (Rosenfeld says: the number of *holes*). If the object B consists of C single (connected) components, we obtain the "phenomenological" formula

$$\psi(B) = \varepsilon - \varkappa + \mu = C - H \quad .$$

By means of Pick's formula, the inequality

$$\varepsilon(C) \leq A(C) + \frac{1}{2}P(C) + 1$$

follows for any convex figure covering a finite part of the orthogonal lattice embedded in the two-dimensional Euclidean plane [No48, Wi90]. Thereby $A(C)$ is the area, and $P(C)$ is the perimeter of the figure.

2.3.4 Planar Semi-Homogeneous Graphs

There are some planar graphs where the requirement of homogeneity is meaningful only for neighborhood degree alone or only for mesh length alone. These graphs are characterized by their construction rules.

The first example are *Poisson graphs*. Here, N straight lines were drawn randomly in that manner that they are pairwise non-parallel and that no more than two lines meet in one point (figure 2.3.-4). Because each line cuts $N-1$ other lines, $\varepsilon=N(N-1)/2$ is the number of points in the corresponding graph.

Otherwise, each line consists of $N-2$ finite line segments and 2 half-rays. Therefore, we have $\varkappa = N(N-2)$ edges in the graph if we neglect all infinite half-rays.

Because the Euler characteristic of the constructed planar graph is $\chi_P = \varepsilon - \varkappa + \omega = 2$, we obtain $\omega = 2 - \varepsilon + \varkappa = 2 + N(N-3)/2$ meshes. The mean neighborhood degree or the mean mesh length are

$$\bar{\nu} = \frac{2\varkappa}{\varepsilon} = 4\,\frac{N-2}{N-1}$$

$$\bar{\lambda} = \frac{2\varkappa}{\mu} = 4\,\frac{N-2}{N-3+\frac{4}{N}} \quad .$$

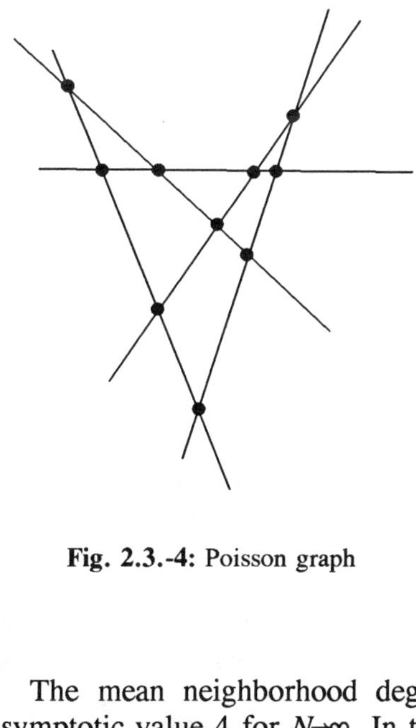

Fig. 2.3.-4: Poisson graph

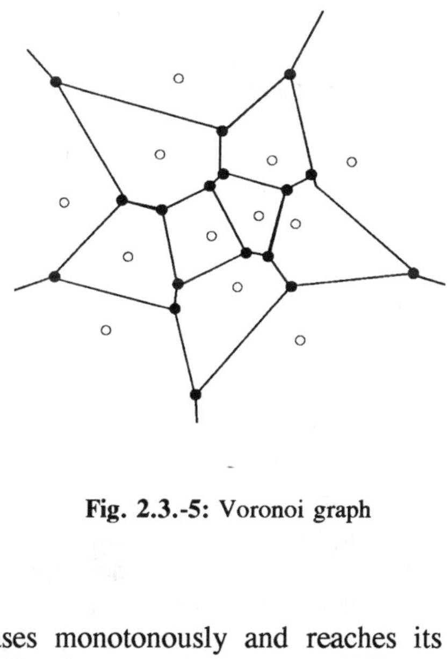

Fig. 2.3.-5: Voronoi graph

The mean neighborhood degree increases monotonously and reaches its asymptotic value 4 for $N \to \infty$. In this limit, all points have exactly 4 neighbors. But we have $\lambda \in \{3,4,\ldots\}$ for the mesh lengths so that the denotation as *semi-homogeneous graph* is justified. The mean mesh lengths reaches the maximal value 4.375 for $N=7$ and goes to the limit 4 for $N \to \infty$.

In figure 2.3.-5, the *Voronoi graph* [Se84, Pr85, Vo88, Vo91] is constructed for P randomly distributed points, the so-called *plane-points*. The Voronoi graph consists of ε vertices, $\varkappa$ edges and ω meshes. We assume that K points of the Voronoi graph are corner points of the convex hull of the set of plane points.

Further, in each of the $\varepsilon - K$ inner vertices of the Voronoi graph only three edges meet each another. If we neglect the infinite long half-rays, only two edges are attached to each of the K outer vertices. Then, the node theorem 1.1.-1 gives

$$\sum \nu(p) = 2K + 3(\varepsilon - K) = 3\varepsilon - K = 2\varkappa \; .$$

Here, $\varkappa$ is the number of finite edges of the Voronoi graph.

Because each of the P plane points lies inside of a mesh of the Voronoi graph and because K unbounded meshes are neglected, we obtain

$$\omega = P - K + 1$$

as mesh number of the Voronoi graph. Using $\chi_V = \varepsilon - \varkappa + \omega = 2$, it follows

$$\begin{aligned} \varepsilon &= 2P - K - 2 \; , \\ \varkappa &= 3P - 2K - 3 \; , \\ \omega &= P - K + 1 \; . \end{aligned}$$

The Voronoi graph is also a semi-homogeneous graph because all neighborhood degrees reaches the value 3 for $P \to \infty$. For finite values of P, we obtain

$$\bar{\nu} = \frac{2\varkappa}{\varepsilon} = \frac{6P - 4K - 6}{2P - K - 2} \; ,$$

$$\bar{\lambda} = \frac{2\varkappa}{\omega} = \frac{6P - 4K - 6}{P - K + 1} \; .$$

It is of interest that $\varepsilon = P-2$, $\varkappa = P-3$, $\omega = 1$ follows if $P = K$, i.e. if all planar points are corner points of their own convex hull. In this case, the Voronoi graph is an unbranched tree.

The last example for semi-homogeneous graphs shall be the *Delaunay graph* as dual graph of the Voronoi graph (see figure 2.3.-6). Here, all meshes have the same length $\lambda = 3$, but the only outer mesh has the length K with K as number of corners of the convex Delaunay hull.

The number of planar points gives the number of points of the Delaunay graph. Finally, with $\chi_D = \varepsilon - \varkappa + \omega = 2$ follows

$$\begin{aligned} \varepsilon &= P \; , \\ \varkappa &= 3P - K - 3 \; , \\ \omega &= 2P - K - 1 \; . \end{aligned}$$

The mean values are

$$\bar{\nu} = \frac{6P - 2K - 6}{P} ,$$

$$\bar{\lambda} = \frac{6P - 2K - 6}{2P - K - 1} .$$

We obtain the value 6 for the mean neighborhood degree of the Delaunay graph in the limit $P \to \infty$, and then all meshes have the length 3.

Because the corresponding values for the Voronoi graph are 3 and 6, respectively, the duality between both graphs is emphasized once more. These three semi-homogeneous graphs are so to speak modifications of the three homogeneous toroidal structures $\Gamma(\nu,\lambda)$ as discussed in section 2.3.2.

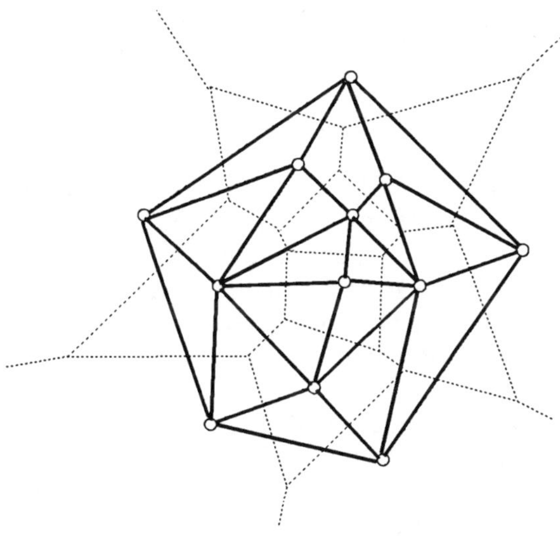

Fig. 2.3.-6: Delaunay graph

2.4 Objects in N-Dimensional Incidence Structures

2.4.1 Three-Dimensional Homogeneous Incidence Structures

In chapter 2.2, we have developed the theory of two-dimensional neighborhood structures as an approach to digital image analysis. Now the results of this theory will generalize for three and more dimensions. Such a theoretical generalization is necessary because spatial image series yielded by computer tomography and confocal laser scan microscopy must be effectively evaluable by exactly defined algorithms.

Theoretical approaches to three-dimensional images have already been investigated in the past [Bi84, Gr71, Le86a, Pa71]. But there is no generally applicable theory for n-dimensional objects in grid point spaces Z^n. However, the mathematical discipline of cellular complexes (section 2.1.2) gives hints how to develop such a theory of n-dimensional images, objects, and surfaces [Ko87, Le86, Ri75]. The basic notion is that of homogeneity of n-dimensional incidence structures. This approach allows a generalization of the theory of two-dimensional neighborhood structures to more than two dimensions.

In Z^2, we know the relationship $1/\nu + 1/\lambda = 1/2$ for the structure constants ν and λ (number of neighbors and length of meshes). We show here in the general case of Z^n that there are $n(n+1)$ structure constants which are however not independent of one another. The "orthogonal" grid space Z^n is one of the most important models for n-dimensional homogeneous incidence structures. For $n=2$ we have three different structure models, and there are eleven different structure models for $n=3$.

An n-dimensional incidence structure $[E,I]$ is a set E of elements as union of $n+1$ disjoint classes $E_0,E_1,\ldots,E_n$ with a reflexive and symmetrical relation I between the elements (see definition 2.1.-3). In a finite structure, a_k is the number of elements in class E_k. For each k-dimensional element $e \epsilon E_k$, there are $b_{kl}(e)$ l-dimensional elements $e' \epsilon E_l$ with $(e,e') \epsilon I$. Especially is $b_{kk}(e)=1$ for all k-dimensional elements.

With respect of the symmetry of incidence relation I, it follows the matching theorem (section 2.1.3). An n-dimensional homogeneous incidence structure is defined by $b_{kl}(e)=b_{kl}=\text{const}$ for all elements $e \epsilon E_k$. Therefore we obtain $a_k b_{kl} = a_l b_{lk}$ and related generalizations (formula 2.1.-1).

In section 2.3.2, it is shown for a given Euler characteristic χ that two-dimensional finite homogeneous incidence structures of any size (i.e. any large number of elements) only exist for vanishing Euler characteristic.

The Euler characteristic of a finite n-dimensional structure is given by

$$\chi^{(n)} = \sum_{k=0}^{n} (-1)^k a_k$$

as we will shown in further sections. Such structures will we investigate also in higher dimensional cases. For homogeneous structures and $\chi^{(n)}=0$, we obtain with $a_k=a_l b_{lk}/b_{kl}$ the relationships

$$\frac{\chi^{(n)}}{a_l} = \sum_{k=0}^{n} (-1)^k \frac{b_{lk}}{b_{kl}} = 0 \qquad \text{for} \quad 0 \leq l \leq n \quad .$$

Using the relationships between the structure constants derived in section 2.1.3, we can show that all of these $n+1$ equations are equivalent to one another. It is a hard work to solve generally the non-linear diophantic equation system given by the matching theorem for any dimension n and any finite value of $\chi^{(n)}$. But for small values of n and small absolute values of $\chi^{(n)}$, the solution is possible by complete enumeration [Vo88].

For $n=1$, there exists only one type of one-dimensional structures with $\chi^{(1)}=0$ where each element $e \epsilon E_0$ has exactly $\nu=2$ neighbored elements. For $n=2$, there are three different types of structures with $\chi^{(2)}=0$. These structures are the networks $\Gamma(\nu,\lambda)$ with $(\nu,\lambda)=(6,3)$, $(4,4)$, $(3,6)$ where each element $e \epsilon E_2$ is bounded by λ points and λ edges (figure 2.1.-1). The network (4,4) corresponds to the common "orthogonal" lattice Z^2.

Further in this way, we obtain for $n=3$ by computer aided enumeration eleven types of structures with $\chi^{(3)}=0$ (see table 2.4.-1). To complete this table, we have to take into consideration $b_{10}=2$, $b_{13}=b_{12}$, $b_{20}=b_{21}$, and $b_{23}=2$.

Table 2.4.-1: Structure constants b_{kl} for three-dimensional structures

type	b_{01}	b_{02}	b_{03}	b_{12}	b_{21}	b_{30}	b_{31}	b_{32}
I	4	6	4	3	3	4	6	4
II					4	8	12	6
III					5	20	30	12
IV	6	12	8	4	3	4	6	4
V					4	8	12	6
VI					5	20	30	12
VII	12	30	20	5	3	4	6	4
VIII					4	8	12	6
IX					5	20	30	12
X	8	12	6	3	3	6	12	8
XI	20	30	12	3	3	12	30	20

There is the question whether finite homogeneous n-dimensional structures do exist in general. For closed (*toroidal*) structures with $\chi^{(n)}=0$ and neighborhood relations like the "orthogonal" grid point space Z^n hold, with respect to section 2.1.5, the general formula

$$b_{kl}^{(n)} = \begin{cases} 2^{l-k} \binom{n-k}{n-l} & \text{for } k<l \\ 1 & \text{for } k=l \\ 2^{k-l} \binom{k}{l} & \text{for } k>l \end{cases}$$

for the structure constants of Z^n (see also [Ro71, Kl83]).

Using these relationships, the matching formulas can be fulfilled. Because we will investigate finite objects, these expressions are fundamental both for finite closed structures and for infinite grid point spaces Z^n. The structure constants for Z^3 corresponds to case V in table 2.4.-1. About ten years ago, Klette had already investigated these relationships:

> "The cells in Z^n are interesting combinatorical objects. Each k-cell contains exactly $2^{k-l}\binom{k}{l}$ l-cells, and each l-cell is contained in exactly $2^{k-l}\binom{n-l}{k-l}$ k-cells of Z^n, for $0 \le l \le k \le n$.
>
> A *grid diagonal* is a straight line segment between two grid points which are included in one n-cell of Z^n, and which are not included in one grid edge. Each n-cell contains exactly $(2^n-n-1)2^{n-1}$ grid diagonals. By three different grid points included in one n-cell of Z^n, a *grid triangle* is defined. Each n-cell contains exactly
>
> $$\binom{2^n}{3} = \frac{1}{3}(2^n - 1)(2^n - 2)2^{n-1}$$
>
> different grid triangles." [Klette, Kl83]

To each point $p \in E_0$ of Z^3, $b_{01}=6$ other points are neighbored by edges. Further, $b_{02}=12$ meshes (or "faces") and $b_{03}=8$ cells are attached to this point p. To each edge of the structure, $b_{12}=4$ meshes and $b_{13}=4$ cells are attached. Each mesh is characterized by $b_{20}=4$ points and $b_{21}=4$ edges. Finally, $b_{30}=8$ points, $b_{31}=12$ edges and $b_{32}=6$ meshes correspond to any single cell $e \in E_3$ because such a cell is a small cube.

Table 2.4.-2: Structure constants b_{kl} for the "orthogonal" grid point space Z^3

b_{kl}	$l = 0$	1	2	3
$k = 0$	1	6	12	8
1	2	1	4	4
2	4	4	1	2
3	8	12	6	1

Box-like parts of Z^n with structure constants given above will be defined as "n-dimensional images". The elementary units of such structures are segments for $n=1$, squares for $n=2$, and cubes for $n=3$.

2.4.2 Objects in Z^n

A finite subset $B \subset E_0$ of grid points of Z^n is called an *object*. The connectivity of an object is determined by a neighborhood relation: two points $p,q \in E_0$ are neighbored if they are attached to the same edge (see section 1.2.2). Each point $p \in B$ is an object point. A k-dimensional element $e \in E_k$ is called an *object element* if all b_{kl} attached l-dimensional elements with $l<k$ are also object elements. This recursive definition is equivalent to the following definition.

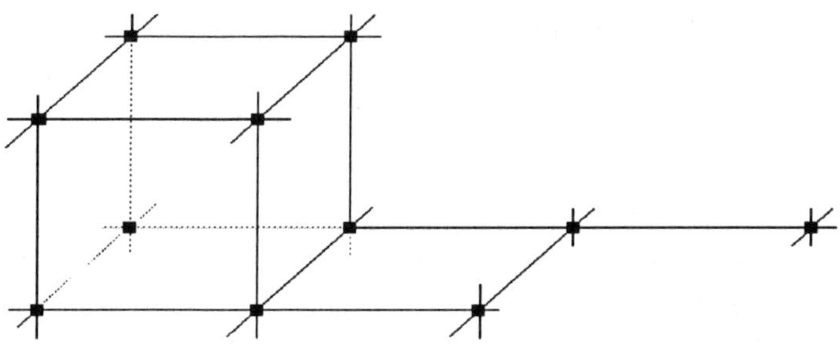

Fig.2.4.-1: This object ($\varepsilon=11$, $\varkappa=16$, $\mu=7$, $\zeta=1$) is characterized by $c_{01}=34$, $c_{12}=36$, and $c_{23}=8$

Def. 2.4.-1: A k-dimensional element $e \in E_k$ is called an *object element* of B if all of its $b_{k0}=2^k$ attached points $p \in E_0$ are object points, i.e. belong to B. An element e' of Z^n is denoted as *marginal element* of an object B if e' is no object element of B but if at least one point of B is attached to e'.

The symbol $c_{lk}(e)$ with $l<k$ denotes the number of k-dimensional marginal elements which are attached to the l-dimensional object element e. The sums

$$c_{lk} = \sum_{e \in E_l} c_{lk}(e)$$

taken for all l-dimensional object elements of B are called *marginal numbers* (see figure 2.4.-1). For an object B, the general matching equations have to be replaced by

Formula 2.4.-1:

$$\begin{aligned} a_l b_{lk} - c_{lk} &= a_k b_{kl} \quad \text{for} \quad k > l \;, \\ a_l b_{lk} &= a_k b_{kl} \quad \text{for} \quad k = l \;, \\ a_l b_{lk} + c_{kl} &= a_k b_{kl} \quad \text{for} \quad k < l \;. \end{aligned}$$

Here a_k means as usual the number of k-dimensional object elements. The first group of equations follows because all l-dimensional elements attached to a k-dimensional element belong likewise to the object B. But there are c_{lk} k-dimensional elements among all k-dimensional elements attached to l-dimensional object elements which do not belong to the object B. Finally, the last group of equations follows by reversing k and l.

These equations - called *object matching formulas* - express the inhomogeneity of objects contrary to the homogeneity of the underlying incidence structure. For any object B of Z^n, we define

$$\psi^{(n)}(B) = \sum_{k=0}^{n} (-1)^k a_k(B)$$

as *Euler number* of the object B. The Euler number of an object can be different from zero as the example in figure 2.4.-1 shows ($\psi^{(3)} = \varepsilon - \varkappa + \mu - \zeta = 1$). A motivation for this alternating sum is given in section 2.4.4 where $\psi = q_0$ is shown to be invariant with respect to object magnification. It is (see formula 2.1.-3)

$$\frac{\binom{l}{k}}{\binom{n-k}{n-l}} = \frac{\binom{n}{k}}{\binom{n}{l}} \quad \text{for} \quad k<l$$

$$\frac{\binom{n-l}{n-k}}{\binom{k}{l}} = \frac{\binom{n}{k}}{\binom{n}{l}} \quad \text{for} \quad k>l \;.$$

Therefore using the structure constants of Z^n, we obtain

$$\sum_{k=0}^{n} (-1)^k \frac{b_{lk}}{b_{kl}} = \sum_{k=0}^{l-1} (-1)^k \frac{b_{lk}}{b_{kl}} + (-1)^l + \sum_{k=l+1}^{n} (-1)^k \frac{b_{lk}}{b_{kl}}$$

$$= \frac{1}{\binom{n}{l}} \sum_{k=0}^{n} (-1)^k \binom{n}{k} = 0$$

for any l with $0 \le l \le n$ corresponding to $\chi^{(n)}=0$ in section 2.4.1 (see also section 2.1.5). With respect to this, the Euler number of an object B can be expressed by the object matching formulas so that

$$\psi^{(n)}(B) = \sum_{k=0}^{l-1} (-1)^k \frac{c_{kl}(B)}{b_{kl}} - \sum_{k=l+1}^{n} (-1)^k \frac{c_{lk}(B)}{b_{kl}}$$

follows for any l with $0 \le l \le n$. This formula means that the Euler number of any object in Z^n can be determined by counting of marginal elements alone (for $n=3$, Lee and Rosenfeld have shown that $\psi^{(3)}$ can be determined by using the Gaussian curvature of the object surface alone [Le86a]). Using the object matching formula, it is possible to express successively all numbers a_k only by the point number a_0 and the numbers $c_{l-1,l}$:

$$a_k = a_0 \frac{b_{0k}}{b_{k0}} - \sum_{l=1}^{k} \frac{c_{l-1,l}}{b_{l-1,l}} \cdot \frac{b_{l-1,k}}{b_{k,l-1}} .$$

This equation is right for $k=0$. Assuming the correctness for a given k, we have for fixed l

$$a_{k+1} = a_k \frac{b_{k,k+1}}{b_{k+1,k}} - \frac{c_{k,k+1}}{b_{k+1,k}}$$

$$= a_0 \frac{b_{0k} b_{k,k+1}}{b_{k+1,k} b_{k0}} - \sum_{l=1}^{k} \frac{c_{l-1,l} b_{l-1,k} b_{k,k+1}}{b_{l-1,l} b_{k,l-1} b_{k+1,k}} - \frac{c_{k,k+1}}{b_{k+1,k}}$$

$$= a_0 \frac{b_{0,k+1}}{b_{k+1,0}} - \sum_{l=1}^{k+1} \frac{c_{l-1,l} b_{l-1,k+1}}{b_{l-1,l} b_{k+1,l-1}}$$

so that the equation is also fulfilled for $k+1$.

Taking into account this general expression for a_k and the formula $\Sigma(-1)^k b_{0k}/b_{k0}=0$, we obtain

$$\psi^{(n)} = \sum_{k=0}^{n} (-1)^k a_k = \sum_{k=0}^{n} (-1)^k \left(- \sum_{l=1}^{k} \frac{c_{l-1,l} b_{l-1,k}}{b_{l-1,l} b_{k,l-1}} \right) .$$

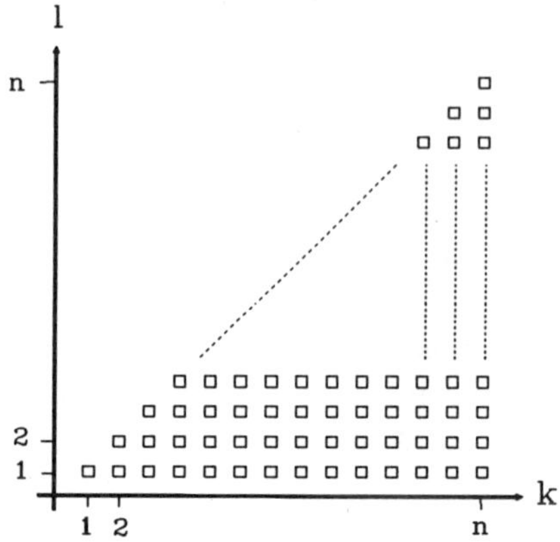

Fig. 2.4.-2: Rearrangement of double sum for k and l

The double sum can be rearranged. Because such a procedure will be used also in the following proofs, it shall be demonstrated in figure 2.4.-2.

We have to determine the sum of all sum terms s_{kl} for a given region in the k-l-plane. The terms to be taken into consideration are symbolized by squares in figure 2.4.-2. The total sum value is independent of the way we select the sum terms s_{kl} for addition.

In this case, we can first sum for l and then for k or in the opposite order. Therefore, it follows by rearrangement of sum terms the formula

$$\psi^{(n)} = \sum_{l=1}^{n} \frac{c_{l-1,l}}{b_{l-1,l}} \sum_{k=l}^{n} (-1)^{k+1} \frac{b_{l-1,k}}{b_{k,l-1}} .$$

for the Euler number. If we consider the structure constants b_{kl} for Z^n, the formulas 2.1.-3, and the identity

Formula 2.4.-2:

$$-\sum_{k=m+1}^{n} (-1)^k \binom{n}{k} = \sum_{k=0}^{m} (-1)^k \binom{n}{k} = (-1)^m \binom{n-1}{m}$$

which is provable by induction from m to $m+1$, then it follows

Formula 2.4.-3:

$$\psi^{(n)} = \frac{1}{2n} \sum_{l=1}^{n} (-1)^{l+1} c_{l-1,l} .$$

This is a very simple formula for the Euler number $\psi^{(n)}$ of an object in Z^n. Independent of dimension n, the $c_{l-1,l}$ are

the number c_{01} of all marginal edges counted for all object points,

the number c_{12} of all marginal meshes counted for all object edges,

the number c_{23} of all marginal cells counted for all object meshes .

All $c_{l-1,l}$ appearing here and in many further formulas can be expressed by the a_k's as the object matching formula shows:

$$c_{l-1,l} = a_{l-1} b_{l-1,l} - a_l b_{l,l-1} \quad .$$

2.4.3 Similarity of Objects

A fundamental notion in geometry is that of *similarity*. Similar objects can be obtained in discrete geometry by magnification of objects or by refinement of the Z^n-lattice. At all coordinate axis between x_l and x_l+1, we introduce new values $x_l' = r \cdot x_l$, $r \cdot x_l+1$, $r \cdot x_l+2, \ldots r \cdot x_l+r-1$. Both values $r \cdot x_l$ and $r \cdot (x_l+1)$ form the support of a "super lattice" in the new lattice (see figure 2.4.-3).

This definition of similarity has the advantage that a point gets a point, a line gets a line, etc. The first to use this notion of magnification of a curve-like object was Freeman [Fr61]:

> "The process of expansion is performed as illustrated by the following example. Given a curve represented by (the code sequence) 012075, a curve exactly twice this size, but otherwise indistinguishable, is given by 001122007755. To expand a curve by a ratio *n*, each of the digits of the curve must be replaced by a set of *n* digits. One notes that *n* must be an integer." [Freeman, Fr61]

By means of *r*-fold magnification, each old *k*-dimensional object element yields a *k*-dimensional *r-cube*, i.e. a cube with $(r+1)^k$ object points. The *l*-dimensional object elements of the magnified object can be parts of different *k*-cubes for $k \geq l$. For instance, the points of the magnified object are caused by points, edges, meshes... of the old object (see figure 2.4.-3).

We will determine the number of *l*-dimensional elements which are caused only by *k*-cubes but not by *m*-cubes with $l \leq m < k$. Each *l*-dimensional element is characterized by an origin $(x_1, \ldots, x_i, \ldots, x_n)$ and by l pairs $\{x_i, x_i+1\}$ which determine the 2^l points of the *l*-dimensional element. Within the *k*-cube, there are r possibilities for the values of each of the l pairs so that both points x_i and

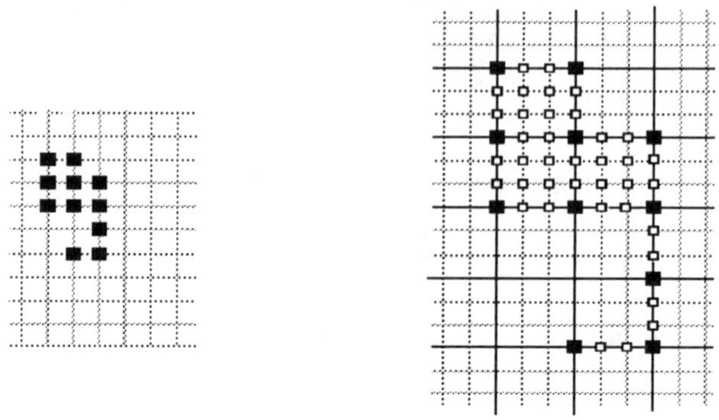

Fig. 2.4.-3: Threefold refinement of a two-dimensional lattice and magnification of an object

x_i+1 lie inside the k-cube. Therefore, we obtain r^l possibilities for l-dimensional elements using l varying coordinates of the element origin and $k-l$ fixed coordinates.

The $k-l$ fixed coordinates can have only $r-1$ different values because the "extremal values" would give l-dimensional elements at the surface of the k-cube. In total, we obtain $r^l(r-1)^{k-l}$ possibilities that a given l-dimensional element lies inside a k-cube. Because we can select any l-tupel of k coordinates, each k-dimensional element e of the old lattice Z^n leads to

$$a_l'(e) = \binom{k}{l} r^l (r-1)^{k-l}$$

l-dimensional elements of the refined lattice for $l \le k$. Therefore the total number a_l' of l-dimensional elements of the magnified object is

$$a_l' = \sum_{k=1}^{n} \sum_{e \in E_k} a_l'(e) = \sum_{k=1}^{n} \binom{k}{l} r^l (r-1)^{k-l} a_k \quad .$$

Especially, we obtain for $n=2$

$$\varepsilon' = \varepsilon + (r-1)\varkappa + (r-1)^2\mu \quad ,$$

$$\varkappa' = r\varkappa + 2r(r-1)\mu \quad ,$$

$$\mu' = r^2\mu \quad ,$$

and for $n=3$

$$\begin{aligned}
\varepsilon' &= \varepsilon + (r-1)\varkappa + (r-1)^2\mu + (r-1)^3\zeta \quad,\\
\varkappa' &= r\varkappa + 2r(r-1)\mu + 3r(r-1)^2\zeta \quad,\\
\mu' &= r^2\mu + 3r^2(r-1)\zeta \quad,\\
\zeta' &= r^3\zeta \quad.
\end{aligned}$$

Using these formulas, one can derive the relationships

$$\begin{aligned}
\varepsilon' - \varkappa' + \mu' &= (\varepsilon - \varkappa + \mu)\,r^0 \quad,\\
\varkappa' - 2\mu' &= (\varkappa - 2\mu)\,r^1 \quad,\\
\mu' &= \mu\,r^2
\end{aligned}$$

for $n=2$, and the relationships

$$\begin{aligned}
\varepsilon' - \varkappa' + \mu' - \zeta' &= (\varepsilon - \varkappa + \mu - \zeta)\,r^0 \quad,\\
\varkappa' - 2\mu' + 3\zeta' &= (\varkappa - 2\mu + 3\zeta)\,r^1 \quad,\\
\mu' - 3\zeta' &= (\mu - 3\zeta)\,r^2 \quad,\\
\zeta' &= \zeta\,r^3
\end{aligned}$$

for $n=3$. It can be seen that there are "dimensionless" characteristics, the Euler numbers $\psi^{(2)}=\varepsilon-\varkappa+\mu$ and $\psi^{(3)}=\varepsilon-\varkappa+\mu-\zeta$, and other characteristics which can be transformed by powers of the magnification factor r. Therefore, a general investigation should be successful.

2.4.4 General Surface Formulas

The terms in the transformation formulas for a_l' can be rearranged in respect to powers of magnification factor r:

$$a_l' = \sum_{m=l}^{n} \binom{m}{l} r^m q_m \ .$$

Here q_m means

Formula 2.4.-4:

$$q_m = \sum_{j=0}^{n-m} (-1)^j \binom{j+m}{j} a_{j+m} \ .$$

We obtain by lattice refinement or object magnification

$$q_m' = \sum_{j=0}^{n-m} (-1)^j \binom{j+m}{j} a_{j+m}' \quad .$$

In consideration of the new a_k's, it follows

$$q_m' = \sum_{j=0}^{n-m} (-1)^j \binom{j+m}{j} \sum_{k=j+m}^{n} \binom{k}{j+m} r^k q_k$$

$$= \sum_{k=m}^{n} \sum_{j=0}^{k-m} (-1)^j \binom{j+m}{j} \binom{k}{j+m} r^k q_k$$

$$= \sum_{k=m}^{n} \binom{k}{m} r^k q_k \sum_{j=0}^{k-m} (-1)^j \binom{k-m}{j} \quad .$$

The second sum has the value 1 for $k=m$ and otherwise the value 0. Because of that, we have $q_m'=r^m q_m$, and therefore the q_m defined above are *object characteristics* of *similarity degree m*. They can be determined by numbers a_k of k-dimensional elements of an object:

$$q_0 = a_0 - a_1 + a_2 - a_3 + a_4 - a_5 \pm \ldots\ldots + (-1)^n a_n \quad ,$$

$$q_1 = a_1 - 2a_2 + 3a_3 - 4a_4 + 5a_5 \mp \ldots\ldots + n(-1)^n a_n \quad ,$$

$$q_2 = a_2 - 3a_3 + 6a_4 - 10a_5 \pm \ldots + \binom{n}{2}(-1)^n a_n \quad ,$$

$$q_3 = a_3 - 4a_4 + 10a_5 \mp \ldots + \binom{n}{3}(-1)^n a_n \quad ,$$

$$q_4 = a_4 - 5a_5 \pm \ldots + \binom{n}{4}(-1)^n a_n \quad .$$

Now, we can take into account the relationships where the a_k are expressed by the numbers $c_{l-1,l}$ of marginal elements. Then, it follows from the a_k as functions of c's and b's

$$q_m = \sum_{j=0}^{n-m} (-1)^j \binom{j+m}{j} a_0 \frac{b_{0,j+m}}{b_{j+m,0}} -$$

$$- \sum_{j=0}^{n-m} (-1)^j \binom{j+m}{j} \sum_{l=1}^{j+m} \frac{c_{l-1,l}}{b_{l-1,l}} \cdot \frac{b_{l-1,j+m}}{b_{j+m,l-1}} \quad .$$

Using the structure constants of Z^n (formula 2.1.-2), we obtain

$$q_m = a_0 \binom{n}{m} \sum_{j=0}^{n-m} (-1)^j \binom{n-m}{j} -$$

$$- \sum_{j=0}^{n-m} (-1)^j \binom{j+m}{j} \sum_{l=1}^{j+m} \frac{c_{l-1,l}}{2} \cdot \frac{\binom{n-(l-1)}{n-(j+m)}}{\binom{n-(l-1)}{n-(j+m)} \binom{j+m}{l-1}} .$$

The first term vanishes for $m<n$. The second term can be transformed to

$$q_m = - \sum_{j=0}^{n-m} (-1)^j \sum_{l=1}^{j+m} \frac{c_{l-1,l}}{2} \cdot \frac{(n-l)!\ (l-1)!}{(n-m)!\ m!} \binom{n-m}{j}$$

$$= - \binom{n}{m} \sum_{j=0}^{n-m} (-1)^j \binom{n-m}{j} \sum_{l=1}^{j+m} \frac{c_{l-1,l}}{2l} \frac{1}{\binom{n}{l}}$$

or by rearrangement

$$q_m = - \binom{n}{m} \sum_{l=1}^{n} \frac{c_{l-1,l}}{2l} \frac{1}{\binom{n}{l}} \sum_{j=0}^{n-m} (-1)^j \binom{n-m}{j} +$$

$$+ \binom{n}{m} \sum_{l=m+1}^{n} \frac{c_{l-1,l}}{2l} \frac{1}{\binom{n}{l}} \sum_{j=0}^{l-m-1} (-1)^j \binom{n-m}{j} .$$

Once more, the first term vanishes for $m<n$. The second term can be transformed using the identity formula 2.4.-2 to

$$q_m = \binom{n}{m} \sum_{l=m+1}^{n} \frac{c_{l-1,l}}{2l} \frac{1}{\binom{n}{l}} (-1)^{l-m-1} \binom{n-m-1}{l-m-1} .$$

Finally, we obtain for $0 \leq m < n$ the simple but very important *geometric main formula*

<u>Formula 2.4.-5:</u>

$$q_m^{(n)} = \frac{1}{2(n-m)} \sum_{l=m+1}^{n} (-1)^{l-m+1} \binom{l-1}{m} c_{l-1,l} .$$

Especially, it follows for the quantities used in image processing:

$$q_0^{(0)} = a_0 = \varepsilon$$

$$q_0^{(1)} = \frac{c_{01}}{2} = \varepsilon - \varkappa = k$$
$$q_1^{(1)} = a_1 = \varkappa$$

$$q_0^{(2)} = \frac{c_{01}-c_{12}}{4} = \varepsilon - \varkappa + \mu = \frac{k-l}{4}$$
$$q_1^{(2)} = \frac{c_{12}}{2} = \varkappa - 2\mu = \frac{l}{2}$$
$$q_2^{(2)} = a_2 = \mu$$

$$q_0^{(3)} = \frac{c_{01}-c_{12}+c_{23}}{6} = \varepsilon - \varkappa + \mu - \zeta$$
$$q_1^{(3)} = \frac{c_{12}-2\,c_{23}}{4} = \varkappa - 2\mu + 3\zeta$$
$$q_2^{(3)} = \frac{c_{23}}{2} = \mu - 3\zeta$$
$$q_3^{(3)} = a_3 = \zeta$$

Here, k is the number of boundary pairs, and l is the contur length of a two-dimensional object.

2.4.5 Interpretation of Object Characteristics

We have shown already in section 2.4.2 that the Euler number $\psi^{(n)}=q_0$ of an object can be determined only by numbers $c_{l-1,l}$ of marginal elements. The marginal elements are attached to the surfaces of the object, and therefore $c^{(i)}{}_{l-1,l}$ means the number $c_{l-1,l}$ for the i-th surface. With

$$\psi^{(n,i)} = \frac{1}{2n} \sum_{l=1}^{n} (-1)^{l+1} c_{l-1,l}^{(i)}$$

it follows that

$$\psi^{(n)} = \sum_{\text{surfaces } i} \psi^{(n,i)} \quad .$$

Similar formulas hold also for the object geometric characteristics q_m, because the geometrical main formula shows that we can define $q_0, q_1, \ldots q_{n-1}$ for single surfaces. For $n=3$, we can determine the Euler number $\psi^{(3,i)} = 1-g^{(i)} = 1-T^{(i)} \in \{1,0,-1,\ldots\}$ depending on the genus $g^{(i)}$ or on the number $T^{(i)}$ of "tunnels" through the i-th surface. Therefore the "phenomenological" formula for the Euler number of three-dimensional object is

$$\psi^{(3)} = \sum_{\text{surfaces } i} (1 - g^{(i)}) = S - T$$

with S as number of surfaces and T as number of tunnels through all these surfaces.

It can be seen from the formulas of section 2.4.3 that there are "dimensionless" characteristics, the Euler numbers $\psi^{(2)} = \varepsilon - \varkappa + \mu$ or $\psi^{(3)} = \varepsilon - \varkappa + \mu - \zeta$. But there exist also "linear" characteristics: $\varkappa - 2\mu = l/2$ for $n=2$ with $l = c_{12}$ as contur length, $\varkappa - 2\mu + 3\zeta = (l-2m)/4$ for $n=3$ with $l = c_{12}$ as number of marginal meshes, and $m = c_{23}$ as number of marginal cells. The expression $\varkappa - 2\mu + 3\zeta$ for three-dimensional objects in Z^3 corresponds to the mean curvature integral [Sa76] for objects in E^3, and $\mu - 3\zeta = m/2$ is a "quadratic" measure for the surface content of an object in Z^3 (Rosenfeld offers the number c_{01} of nodes of the surface graph as measure for "surface area" [Ro89]).

Because the q_m's have to be transformed by the factor r^m, we can derive a lot of "nonlinear" dimensionless numbers. For two-dimensional objects in Z^2, the *shape factor* or *form factor*

Formula 2.4.-6:

$$f = \frac{q_1^2}{4\,q_2} = \frac{(\varkappa - 2\mu)^2}{4\mu} = \frac{l^2}{16\mu} \geq 1$$

exists for $\mu > 0$, which has the same value for all similar objects, and reaches its minimal value for square-shaped objects. The proof for this statement can be given in the following way:

1 - Let X be any connected object with μ meshes. Let $F(X)$ be the corresponding "filled" object where the included holes are filled with points, edges, and meshes. It is

$$\frac{c_{12}^{(a)}}{2} = q_1(F) \leq q_1(X) = \frac{c_{12}(X)}{2} = \frac{c_{12}^{(a)} + \sum c_{12}^{(i)}}{2},$$

$$\mu^{(a)} = q_2(F) \geq q_2(X) = \mu^{(a)} - \sum \mu^{(i)}$$

where the superscripts (a) denote values for the outer boundary and values for the inner boundaries, respectively. Therefore we obtain $f(F)=q_1^2(F)/q_2(F)\leq q_1^2(X)/q_2(X)=f(X)$.

2 - Let F be any connected object without holes with μ meshes. Let $R(F)$ be the *circumscribing rectangle*, i.e. the maximal object with the same extremal coordinate values as F. It is

$$c_{12}(R) = l(R) = 2\,q_1(R) \leq 2\,q_1(F) = c_{12}(F) = l(F) ,$$

$$\mu(R) = q_2(R) \geq q_2(F) = \mu(F) .$$

It follows $f(R)\leq f(F)$ and therefore $f(R)\leq f(X)$.

3 - Let R be any rectangular shaped object with $M\cdot N$ points. We obtain

$$\varkappa = N(M-1) + M(N-1) ,$$

$$\mu = q_2 = (N-1)(M-1) ,$$

$$q_1 = \varkappa - 2\mu = N + M - 2 .$$

Without restriction, we can assume that $M=N+k\geq N$ with $k\geq 0$. Then the inequality $(2N+k-2)^2\geq 4(N-1)(N+k-1)$ is fulfilled and finally

$$f(R) = \frac{(N+M-2)^2}{4(N-1)(M-1)} \geq \frac{(N+N-2)^2}{4(N-1)^2} = f(S) = 1$$

follows with $f(S)$ as form factor of a square S. Because of that, the inequality $f(X)\geq f(S)=1$ is fulfilled for any connected object X.

In Z^3, there exist two independent shape factors for objects:

<u>Formula 2.4.-7:</u>

$$g = \frac{q_1^2}{3\,q_2} = \frac{(\varkappa - 2\mu + 3\zeta)^2}{3(\mu - 3\zeta)} = \frac{(l - 2m)^2}{24m} \geq 1 ,$$

$$h = \frac{q_1 q_2}{9\,q_3} = \frac{(\varkappa - 2\mu + 3\zeta)(\mu - 3\zeta)}{9\zeta} = \frac{(l - 2m)\,m}{72\zeta} \geq 1 .$$

They have the same numerical values for all similar objects assuming that $\mu>3\zeta$ and $\zeta>0$. The minimal values will be reached for cube-like objects as can be proven with the same proving method as for shape factor f in Z^2.

The quantities q_k can be compared with Minkowski's *quermass-integrals* $M_k(C)$ which are defined for convex sets C in the n-dimensional Euclidean space [Le80]:

Theorem 2.4.-1: The n-dimensional volume of the outer parallel set C_d of a convex set C with distance d between C and C_d is a linear combination of the quermasses $M_k(C)$. The quermasses are defined by the formula

$$V(C_d) = \sum_{k=0}^{n} \binom{n}{k} M_k(C)\, d^k \ .$$

In figure 2.4.-4, a parallel set C_d of the two-dimensional convex set C is represented. The area $A(C_d)$ as two-dimensional volume is given by

$$A(C_d) = A + P \cdot d + \pi \cdot d^2$$

with $A=M_0(C)$ as area of the convex set C and $P=2M_1(C)$ as its perimeter.

For three-dimensional convex sets C, we obtain the formula

$$V(C_d) = V + S \cdot d + M \cdot d^2 + \frac{4\pi}{3} \cdot d^3$$

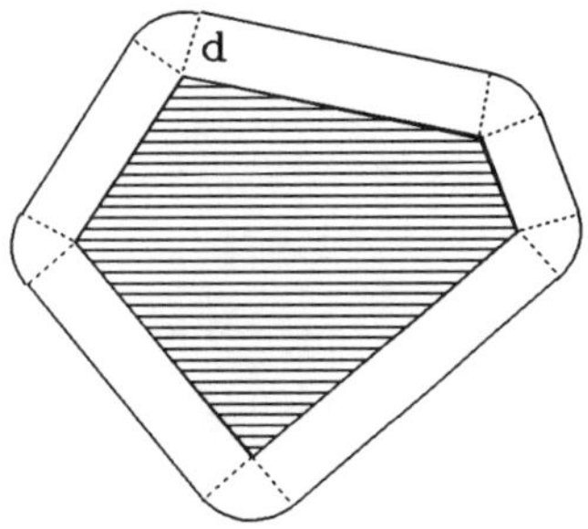

Fig. 2.4.-4: Parallel set C_d of the two-dimensional convex set C

with V as volume, S as surface area, and M as integral of mean curvature of the set C.

In Z^n, the quermass-integrals were investigated by Bieri [Bi84, Bi87] especially in respect to time effective computation of the values of quermass-integrals of objects in binary images. Because uniform regions can be grouped into blocks and the quermass-integrals of blocks can be computed relatively simply, Bieri presents an algorithm using bintree structures in Z^n.

There are some deep analogies between quermass-integrals and our quantities q_k and $c_{l-1,l}$ as it can be shown by investigation of boxes in Z^n. The boxes are so to speak the "convex bodies" of the n-dimensional orthogonal lattice (figure 2.4.-5). The two-dimensional "volume" $q_2=\mu$ fulfills the relationship

$$\begin{aligned}\mu(C_d) &= \mu + 2(l_1 + l_2) \cdot d + 4 \cdot d^2 \\ &= q_2 + 2 \cdot q_1 \cdot d + 4 \cdot q_0 \cdot d^2 \ .\end{aligned}$$

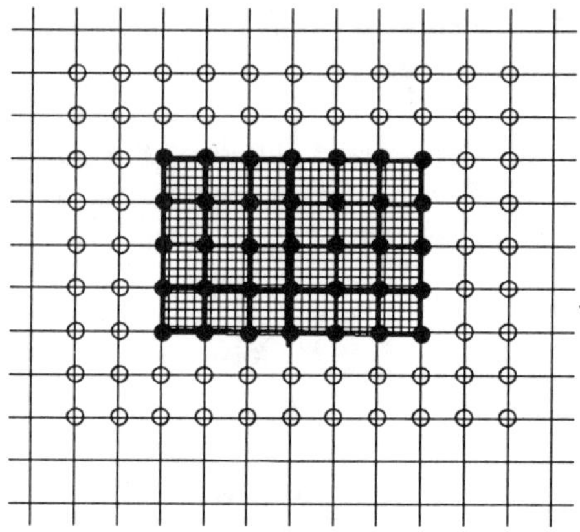

Fig. 2.4.-5: "Convex" object and *parallel body* in the two-dimensional lattice for $d=2$

In the three-dimensional case, we use as "volume" the number ζ of cells. With l_1,l_2,l_3 as "side lengths" of the box, we obtain $\zeta=l_1l_2l_3$ and

$$\begin{aligned}\zeta_d &= (l_1+2d)(l_2+2d)(l_3+2d)\\ &= l_1l_2l_3 + 2(l_1l_2+l_2l_3+l_3l_1)d\\ &\quad + 4(l_1+l_2+l_3)d^2 + 8d^3 .\end{aligned}$$

The expression $2(l_1l_2+l_2l_3+l_3l_1)$ is equal to the number c_{23} of the meshes at the surface of the box, and the term $4(l_1+l_2+l_3)$ is equal to the total edge length of the box, i.e. to the expression $c_{12}-2c_{23}$. With $q_0=\psi^{(3)}=S-T=1$, we obtain

$$\zeta_d = q_3 + 2q_2^{(3)}d + 4q_1^{(3)}d^2 + 8q_0^{(3)}d^3 .$$

Therefore, it would be possible to define the quantities q_k as "discrete quermasses" by means of the "volume" of "parallel bodies" of "convex bodies" in Z^n. But the definition of the q_k's using the similarity of objects (section 2.4.3) is more fruitful because the q_k's can be introduced directly for any objects, i.e. for "convex" and "non-convex" sets of object points likewise.

3. Topological Laws and Properties

3.1 Objects and Surfaces

3.1.1 Surfaces in Discrete Spaces

In section 1.4.2, we have mentioned Clifford's reflections about the boundary of a point set in a discrete space. Boundaries are very important because they separate different point sets, i.e. the local spatial properties change greatly at boundaries. Further, some topological and geometrical characteristics of objects in discrete spaces can be expressed by sum terms corresponding to single surfaces as the formulas of section 2.4.4 show.

In n-dimensional Euclidian spaces, the boundaries are $(n-1)$-dimensional surfaces of objects. For smooth surfaces, a one-dimensional straight line segment is defined for each boundary point (normal vector). A set of local surface normal vectors is called a *needle map* [Wo80, Ik81, Ik83, Ji88, Ji90].

We will use such needle maps also for our theoretical investigations in the discrete spaces Z^n. For that, an appropriate model of a *needle* is given by a *boundary pair*, that means a pair (p,q) of neighbored points with p as object point and q as non-object point (background point). This approach has already been known for many years:

> "If we regard a border as a set of pairs, each consisting of an object point and an adjacent non-object point, we can develop a theory of sequences of such pairs (*bicurves*) analoguous to the theory of curves. In fact, bicurves turn out to be the same as *connected components* of pairs" [Rosenfeld and Morgenthaler, Ro80, Mo81a].

But this model was not consistently carried on in the following time. The temptation to use as many properties of Euclidian geometry as possible also for discrete spaces was too strong:

> "Herman and his colleagues were developing such a surface tracking algorithm...However the surface traced by their algorithm is not a surface in the strict topological sense...Thus it was natural to seek for some closer connections between the two subjects, the newly developed 3D digital topology and the well established Euclidian topology, so that one could use known facts from this latter discipline." [Rosenfeld, Ro85]

"The natural topology on the lattice points is the discrete topology in which every lattice point is an open set; in fact, this is the only nontrivial topology if we impose the natural requirement that open sets go into open sets under (integer) translation. This topology, unfortunately, is not very interesting; for example, the concept of connectedness becomes trivial in this topology, since a set containing more than one lattice point cannot be connected." [Rosenfeld and Melter, Ro87]

"In three dimensions, digital geometry deals with properties of sets of *voxels*; these are upright unit cubes whose vertices have integer coordinates...The *surface* of a set of voxels is composed of *surface pixels*; these are unit squares parallel to the coordinate planes and whose vertices have integer coordinates." [Rosenfeld and Kong, Ro89]

On this base, the Medical Image Processing Group of the University of Pennsylvania has very systematically investigated the three-dimensional surface detection [Ar81, Ar81a, Ud82, He83, Ch85, Ud87, Go89]. Because the basic idea is space construction by voxels, one calls the MIPG-method *cuberille method*. The cuberille method is relatively complicated, since the six faces and the twelve edges of each voxel, depending on their orientations, have to be taken into account individually for the corresponding surface detection algorithm [Ka89].

Contrary to these Euclidian approaches, we will develop here a consistent surface detection algorithm of objects in incidence structures as given by Z^n. The basic idea is the model of a surface as a connected component of a set of boundary pairs [Ro80, Vo90].

3.1.2 Contur Following as Two-Dimensional Boundary Detection

A careful investigation of the logical structure of classical two-dimensional contur following shows that the contur points are not the essential but the boundary pairs outgoing from contur points.

A contur detection starts with a boundary point $p \in B \subset E_0$, that means with an object point neighbored to at least one non-object point q. The pair (p,q) is the first boundary pair of the contur.

In respect of this first boundary pair (p,q), we search neighbored boundary pairs whereby a boundary pair is then neighbored to (p,q) if it can be determined by the algorithm 3.1.-1 represented in figure 3.1.-1. If (p,q) lies in direction u (as example in x-direction) then we have to search in all directions v orthogonal to u (in this case $(+y)$-direction and $(-y)$-direction).

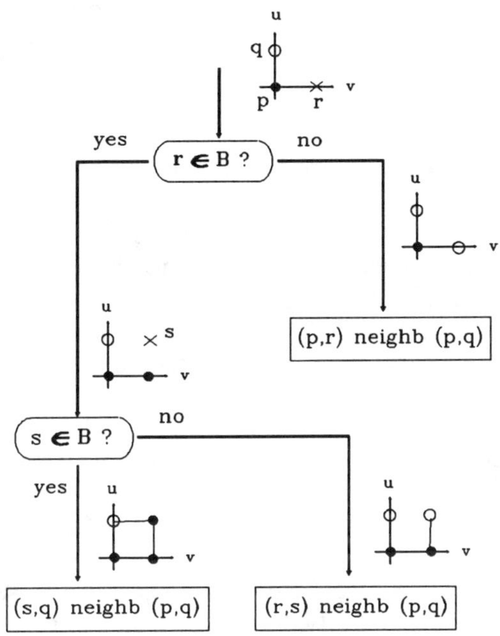

Fig. 3.1.-1: Neighbored boundary pairs with respect to the boundary pair (p,q)

Because the algorithm 3.1.-1 gives for any direction v of the boundary pair (p,q) and for all possible memberships (i.e. $r \in B$ or $r \notin B$ etc.) always a boundary pair neighbored to (p,q), the boundary pair (p,q) has exactly two neighbors.

Therefore, all boundary pairs in the two-dimensional incidence structure Z^2 are neighbored to exactly two other boundary pairs. Because of this fact, the set P of all boundary pairs of an object B is a homogeneous neighborhood structure $[P,N]$ with a neighborhood relation N determined by the algorithm 3.1.-1.

The neighborhood degree of all elements of this neighborhood structure is $\nu=2$. As shown in figure 2.1.-3, a homogeneous neighborhood structure with $\nu=2$ is a one-dimensional ring-like incidence structure. In this sense, the connected components of $[P,N]$ are always "closed curves". The point sequence determined by the object points of successive boundary pairs of one connected component is then a *boundary path* of the object.

An object can have more than one boundary path. The only boundary path with positive total curvature is denoted as *contur* of the object.

As for each neighborhood structure, one can determine the connected components of $[P,N]$ by the general algorithm 1.3.-1 given in section 1.3.2. The algorithm 1.3.-1 needs list processing. But in Z^2, it is possible to dispense with list processing: We can search for the next neighbored boundary pair in clockwise changed direction as successor, and we can be sure that we will finally reach again the start boundary pair.

3.1.3 Three-Dimensional Surface Detection

In case of Z^3, there are four orthogonal directions v (as example $+x,-x,+z,-z$) which correspond to a given direction u (here $+y$ or $-y$). Therefore, each boundary pair has exactly four neighbored boundary pairs determined by the algorithm given in figure 3.1.-1. The connected components of this neighborhood structure of boundary pairs represent the surfaces of a given object B [Ro89].

> "It seems evident that the boundary is, to a certain extent, connected in nature. That is, the location of a boundary element should be connected to the location of the nearest boundary element." [Liu, Li77]

In figure 3.1.-2 a simple three-dimensional object is represented consisting of $\varepsilon=12$ object points, $\varkappa=17$ object edges, $\mu=7$ object meshes, and $\zeta=1$ object cell. This object has 38 boundary pairs which are consecutively numbered.

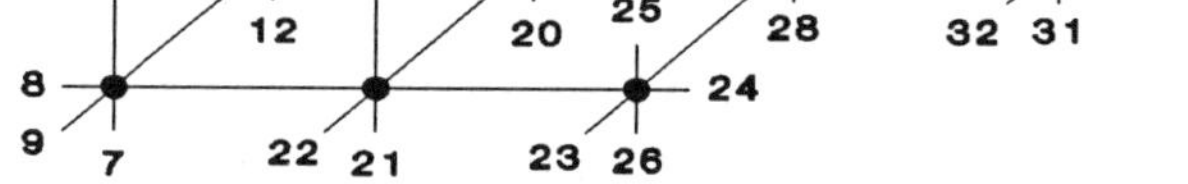

Fig. 3.1.-2: Three-dimensional object with 38 boundary pairs

The neighborhood structure (neighborhood graph) of these 38 boundary pairs is characterized by a neighborhood degree $\nu=4$ (see figure 3.1.-5).

Therefore, the surface detection is a special case of the general graph search algorithm 1.3.-1. The following steps have to be taken to detect a single surface of object B:

Alg. 3.1.-2: Three-dimensional surface detection

1. Determine a not yet unmarked boundary pair $k_0=(p,q)$ with $p \epsilon B$.
2. Initialize a list L of boundary pairs, put the boundary pair k_0 into this list, and mark the boundary pair k_0.
3. Take a boundary pair k from the boundary list L and remove k from L. Prove for each of the four boundary pairs k_1,k_2,k_3,k_4 neighbored to k whether they are marked or not. If k_i is not marked then k_i has to be marked and put into the boundary list L.
4. If the boundary list is non-empty then go to step 3. Otherwise, the surface corresponding to boundary pair k_0 is detected.

The boundary pairs neighbored to a given boundary pair (p,q) can be determined effectively because we know the coordinates of points p and q in the three-dimensional digital image and therefore the coordinates of the points of neighbored pairs, too (see algorithms 1.3.-2/3). But we have to develop another approach for marking boundary pairs to avoid the ineffective use of marking lists.

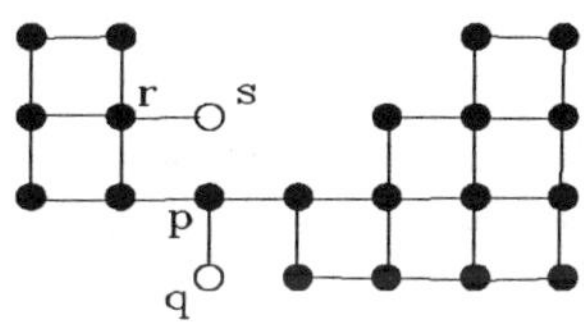

Fig. 3.1.-3: Marking of boundary pairs (see text)

In general, the necessary marking of boundary pairs (p,q) cannot be done by marking of points p and q. Let us assume that the boundary pairs (p,q) and (r,s) in figure 3.1.-3 are already marked by marking of their points. A following test whether the boundary pair (p,s) is already marked then always yields the answer "yes". But this answer is not necessary right.

These difficulties can be avoided by a generalized marking technique: In the n-dimensional lattice Z^n, n edges $k_i=(p,q_i)$ can be unambiguously attached to each point $p=(x_1,x_2,\ldots,x_n)$ whereby the edges go in the positive coordinate direction (as example, $k_1=(p,q_1)$ with $q_1=(x_1+1,x_2,\ldots,x_n)$). In respect of that, each edge is attached to that of its end point which has the smaller coordinate value.

Using now one marking bit for each of the n edges attached to a point, we can "mark an edge" k or "mark a boundary pair" k by setting the marking bit corresponding to this edge in the mark value of the point attached to k. To detect surfaces in n-dimensional images, we have to work therefore with 2^n-valued marks at the lattice points.

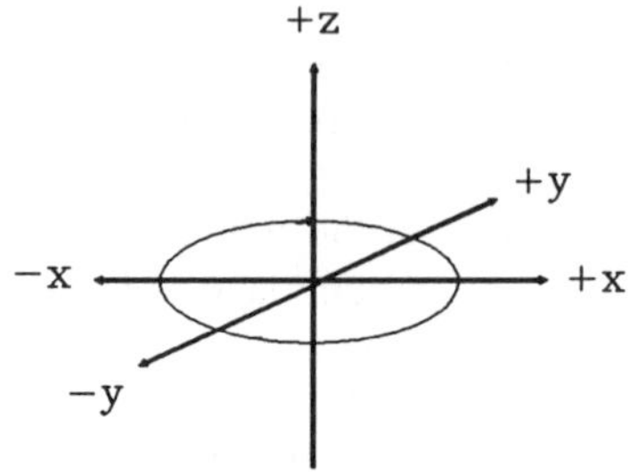

Fig. 3.1.-4: Cyclic order of neighbored boundary pairs

We have pointed out that the boundary pairs of a three-dimensional object represent a neighborhood structure. Further, we can arrange cyclically the four orthogonal directions *v* attached to each direction *u* (see figure 3.1.-4). Therefore, it is possible to arrange cyclically the neighbored boundary pairs with respect to these direction cycles. Thus, an oriented neighborhood structure arises.

In figure 3.1.-5, the only two-dimensional "surface" of the object shown in figure 3.1.-2 is represented as a two-dimensional oriented neighborhood structure Σ_S. In this sense, each surface is a connected component of the boundary pair neighborhood structure.

Because this structure is planar, its Euler characteristic $\chi=\varepsilon-\varkappa+\omega=38-76+40$ has the value 2. Otherwise, the corresponding Euler number $\psi=\varepsilon-\varkappa+\mu-\zeta=12-17+7-1$ of the object of figure 3.1.-2 has the value 1. This is an example for the *Poincare duality* $\chi_{\text{surface}}=2\psi_{\text{object}}$ in three-dimensional spaces [Le86a, Le87].

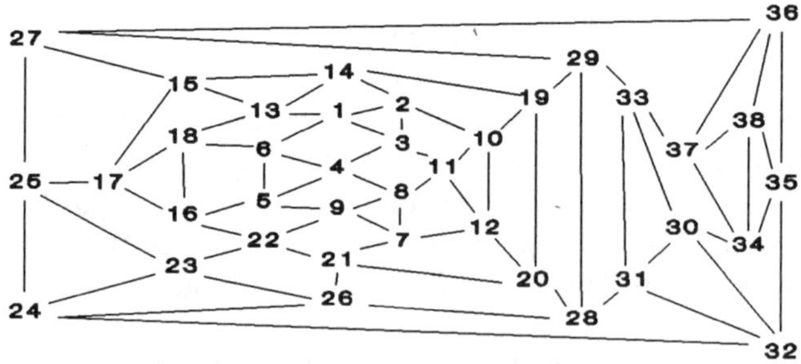

Fig. 3.1.-5: Oriented neighborhood structure Σ_S constructed with the 38 boundary pairs of figure 3.1.-2

It is worth to mention that the algorithm 3.1.-2 is applicable also for object surfaces in Z^n. We have only to replace the four neighbored boundary pairs k_1,k_2,k_3,k_4 by $2n-2$ neighbored boundary pairs. The boundary pairs (p,q) of an $(n-1)$-dimensional surface lying in a selected direction u or in the opposite direction $-u$ open the possibility to determine *input points* and *output points* and therefore also moments attached to this surface (see section 5.5.5).

3.1.4 Curvature of Conturs and Surfaces

The described surface detection algorithm gives the possibility to determine some surface characteristics which are mentioned already in section 2.4.3. The number $k=c_{01}$ is equal to the number of boundary pairs and therefore easily computable.

The number $l=c_{12}$ of marginal edges can be determined also during the surface detection. If the question $s \epsilon B$? in the algorithm of figure 3.1.-1 has the answer "yes", we have to increase l by 2 because the object edges (p,r) and (r,s) have in each case one marginal mesh in respect of direction u. For answer "no", only the object edge (p,r) yields a marginal edge so that l increases by 1.

We can measure the "curvature" of one-dimensional boundaries of two-dimensional objects by the following method. With respect to figure 3.1.-1, a curvature measure c and a number d of object edges are given for each case of neighbored boundary pairs:

case	curvature c	edges d
(p,r)	+1	0
(r,s)	0	1
(s,q)	-1	2

Therefore, we have $c_i=1-d_i$ for each direction of the i-th boundary edge (p,q) or $t_i=2-k_{i,r}-k_{i,l}$ for the "total" curvature t_i of the i-th boundary edge in respect of the numbers $k_{i,r}$ of object edges on the right and $k_{i,l}$ of object edges on the left. The total curvature t of a boundary is therefore the half of the sum over all values t_i:

$$t = \frac{1}{2}\sum_i t_i = \frac{1}{2}\sum_i (2-k_{i,r}-k_{i,l}) = k - l \quad .$$

Here, $k=c_{01}$ means the number of all boundary edges of the boundary, and $l=c_{12}$ is the number of all object edges attached to a boundary mesh of the boundary.

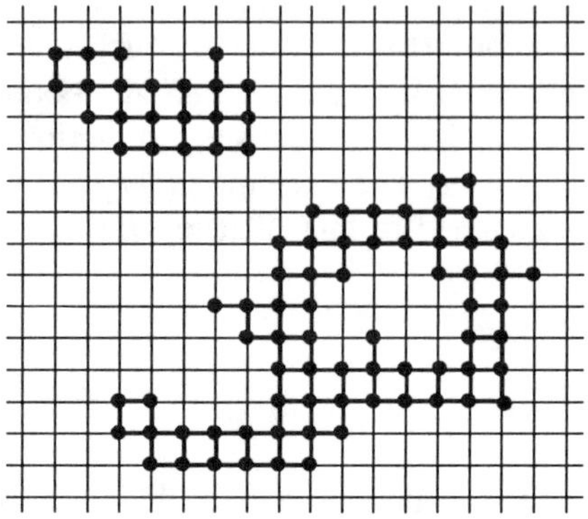

Fig. 3.1.-6: Objects and conturs with curvature $t=k-l$ (see text)

The *total curvature t* of a boundary of a two-dimensional object can have only two values, +4 for an outer boundary (contur of an object), and −4 for an inner boundary (contur of a hole). In figure 3.1.-6, some examples for this law are given:

$$t_a = 24 - 20 = +4 \,,$$
$$t_a = 54 - 50 = +4 \,,$$
$$t_i = 16 - 20 = -4 \,.$$

For surfaces of three-dimensional objects, the determination of $m=c_{23}$ can be obtained also during the process of surface detection, but it is very more complicated.

We obtain for the Euler characteristic of a closed surface (see figure 3.1.-5) the expression

$$\chi_S = \varepsilon_S - \varkappa_S + \omega_S$$

with $\varepsilon_S=c_{01}$ as number of boundary pairs of this surface, whereas $\varkappa_S$ and ω_S cannot be directly determined by marginal numbers c_{kl}. But the surface is characterized by a two-dimensional oriented neighborhood structure Σ_S so that we can use the node theorem 1.1.-1

$$\sum_{p\in P} \nu(p) = 4\varepsilon_S = 2\varkappa_S$$

because in figure 3.1.-5 all points have exactly 4 neighbors. The mesh theorem 2.1.-1 yields

$$\sum_{m\in M} \lambda(m) = 2\varkappa$$

and therefore, we obtain

$$4\chi_S = (4\varepsilon_S - 2\varkappa_S) + (4\omega_S - 2\varkappa_S) = \sum_{\lambda=3} (4\omega_\lambda - \lambda\omega_\lambda)$$

$$= \omega_3 - \omega_5 - 2\omega_6 - 3\omega_7 - \ldots$$

where ω_λ means the number of meshes with length λ in the surface neighborhood structure Σ_S. The mesh lengths λ of the four meshes attached to each point of Σ_S correspond to the configurations of the four marginal cubes attached to the corresponding boundary pair (see figure 3.1.-7).

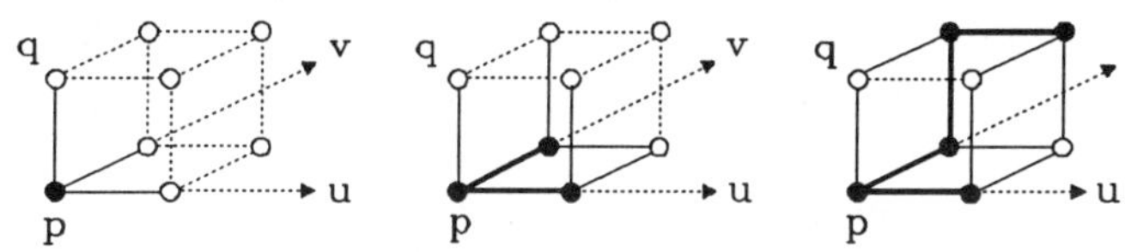

Fig. 3.1.-7: Marginal cubes whose configurations yield meshes of $\sum_S$ with $\lambda=3$ (left), $\lambda=5$ (middle), $\lambda=7$ (right)

During the surface detection, it is therefore possible to determine the numbers ω_λ by investigation of marginal cubes, and therefore we can compute the Euler characteristic of the surface.

Another possibility to determine $\chi(\Sigma_S)$ is given by a special search algorithm for two-dimensional neighborhood structures (see section 2.2.4). Instead of LIFO- or FIFO-technique, we can search successively for nodes of single meshes. For that it is necessary to go to the next node of the graph determined by the orientation of the neighborhood structure. The clockwise arrangement of search directions is shown in figure 3.1.-8.

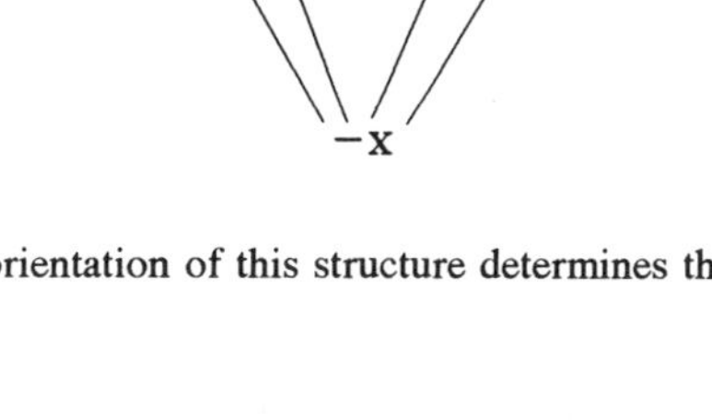

Fig. 3.1.-8: The orientation of this structure determines the orientation of $\sum_S$

For instance, we have found a boundary edge (p,q) in direction $+x$ and search for the next boundary edge (p',q') in direction $-z$. Outgoing from (p',q'), then we must search in direction $+y$ because the neighborhood cycle of the direction $-z$ is given by $c_{-z} = \langle +x,+y,-x,-y \rangle$.

The Euler characteristic of an $(n-1)$-dimensional surface can be used for a decision between an "inner surface" and an "outer surface" only if n is even. In a ring-like structure with $n=1$, there is no "in" or "out" for a connected object. For $n=2$, the curvature of a surface tells us whether we have determined an inner or an outer boundary.

But in the three-dimensional case, an object p consisting of only one point and an object q equal to the whole structure except this one point p have the same Euler number. Therefore the Euler characteristic of the corresponding surfaces are the same although we would speak of an "outer surface" in the first case and an "inner surface" in the second case, respectively.

Generally, these two cases yield for n-dimensional configurations

$$\psi^{(n)}(p) = \sum_{k=0}^{n} (-1)^k a_k = 1 ,$$

$$\psi^{(n)}(q) = \sum_{k=0}^{n} (-1)^k a_k = \sum_{k=0}^{n} \left[(-1)^k a_k^{total} - (-1)^k b_0 k \right]$$

$$= -\sum_{k=0}^{n} (-1)^k 2^k \binom{n}{k} = (-1)^{n+1} .$$

It follows for odd values of n that the Euler number of a one-point-object p is equal to the Euler number of a one-point-hole q. This fact can be compared with the fact that in Euclidean geometry the total curvature of an n-dimensional spherical surface as integral over the product of all main curvatures k_i is

$$T = \int k_1 k_2 \ldots k_{n-1} \, dS \geq 0 ,$$

i.e. always positive if n is odd.

3.2 Motions and Intersections

3.2.1 Motions of Objects in Z^n

In Euclidian geometry, a one-to-one mapping of E^n to E^n is named *motion* if the distance $d(p,q)$ between any two points $p,q \in E^n$ is invariant with respect to this mapping. The three main groups of motions in E^n are translations (addition of a translation vector), rotations (multiplication by an orthogonal matrix with determinant +1), and reflections (multiplication by an orthogonal matrix with determinant -1). Therefore, the analytical expression for a general motion is

$$p' = \begin{pmatrix} x_1' \\ x_2' \\ \vdots \\ x_n' \end{pmatrix} = \begin{pmatrix} r_{11} & r_{12} & \cdots & r_{1n} \\ r_{21} & r_{22} & \cdots & r_{2n} \\ \vdots & \vdots & \vdots & \vdots \\ r_{n1} & r_{n2} & \cdots & r_{nn} \end{pmatrix} \begin{pmatrix} x_1 \\ x_2 \\ \vdots \\ x_n \end{pmatrix} + \begin{pmatrix} s_1 \\ s_2 \\ \vdots \\ s_n \end{pmatrix}$$

where the r_{ik} describe an orthogonal matrix.

It is known that there are only one-to-one "rotations" with multiple values of $\pm 90^0$ in two-dimensional image processing in Z^2:

> "The epsilon-delta definition of continuity has a natural analog for functions that take lattice points into lattice points. It turns out that a function f is 'continuous' if and only if it takes neighbors into neighbors, i.e. if Q is a neighbor of P, then $f(Q)=f(P)$ or is a neighbor of $f(P)$ (diagonal neighbors not allowed)...a function is one-to-one and continuous if and only if it is a combination of translation, rotations by multiples of 90^0, or reflections in a horizontal, vertical, or diagonal line." [Rosenfeld, Ro85]

The corresponding orthogonal matrix in Z^n arises on the one hand from the unit matrix by changing rows or columns. An even number of changes yields the determinant value +1, and an odd number the value -1. On the other hand, we can replace any +1 in the unit matrix by -1 to change the direction of the corresponding coordinate axis (figure 3.2.-1).

In Z^n, we have 2^n possibilities to choose positive or negative signs, and there are $n!$ possibilities to change the rows of the $n \cdot n$-matrix. In half of all these cases, the determinant of the matrix is +1 (in the other half the value is -1). Therefore we obtain $N(n)=2^{n-1}n!$ proper rotations (determinant +1) which we will denote as *tiltings*.

Any object B of Z^n can have $N(n)$ different orientations in Z^n because the coordinate system fixed in the object can be oriented in $N(n)$ ways with respect to the coordinate system fixed in Z^n. In comparison to the tiltings, the translations of an object are relatively simple because we can shift the object in any coordinate direction by any integer number of steps.

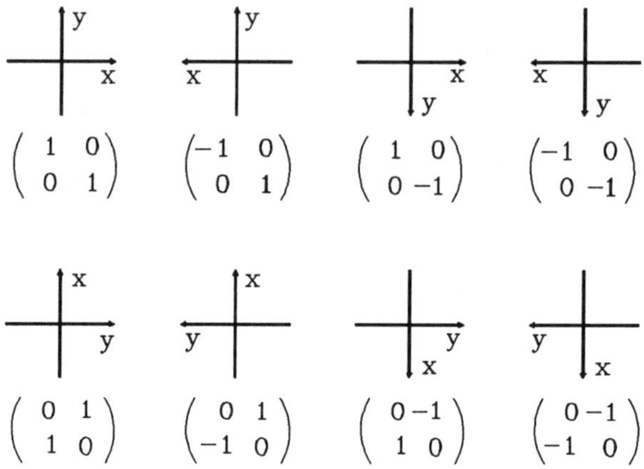

Fig. 3.2.-1: Rotations and reflections in Z^2

<u>Def. 3.2.-1:</u> An integer translation or an integer rotation (tilting with determinant +1) of an object B in Z^n determined by translation or rotation of the object fixed coordinate system is denoted as *motion* of the object.

A k-dimensional object element e of an object B (for instance a neighborhood pair of object points, see definition 2.4.-1) can be interpreted as a special object. How many tiltings are in Z^n which map such an object e onto itself?

Now, a k-dimensional element determines a k-dimensional sub-space of Z^n. In this sub-space, all $2^k k!$ proper and non-proper rotations of Z^k map e onto itself. On the other hand, all $2^{n-k}(n-k)!$ proper and non-proper rotations of the complementary $(n-k)$-dimensional sub-space have no influence of the orientation of e. Out of these $2^k k! \cdot 2^{n-k}(n-k)!$ possibilities, only such mappings are allowed in Z^n which are characterized by determinant +1. Therefore the number of self-mapping tiltings in respect of a k-dimensional element is

$$N_k^{(n)} = 2^{n-1}\, k!\, (n-k)!\ .$$

3.2.2 Count Measures and Intersections of Objects

In chapter 2.4, we have introduced quantities of objects which characterize topological and geometrical properties of these objects:

a_k number of k-dimensional object elements, for instance
- ε number of points of an object B
- $\varkappa$ number of object edges of B
- μ number of object meshes of B (for $n>1$)
- ζ number of object cells of B (for $n>2$)

c_{kl} marginal numbers (see section 2.4.2)

$q_m^{(n)}$ object characteristics of similiarity degree m (especially, q_0 is the Euler number of the object, see section 2.4.4)

These quantities can be determined by counting because all objects to be investigated and all n-dimensional images to be taken into account are characterized by finite sets. Therefore we will call these quantities *count measures.*

It is very important for us that the a_k's determine definitely all other count measures:

$$c_{kl} = a_k b_{kl} - a_l b_{lk} ,$$

$$q_m^{(n)} = \sum_{j=0}^{n-m} (-1)^j \binom{j+m}{j} a_{j+m} .$$

Therefore, it is possible from the theoretical standpoint that we have only to determine the a_k's although the determination of the marginal numbers c_{kl} can be done more effectively using the algorithm of n-dimensional surface detection.

The determination of $a_0=\varepsilon$ as number of points of an object B is very simple: We have to move a test point through that n-dimensional box which contains B, and have to count how many times the test point hits an object point of B.

The same method can be used to determine the number $a_1=\varkappa$ of object edges of B. But "moving of a test edge" means not only translations but also rotations of the test edge. We generalize the counting method by the following theorem.

Theorem 3.2.-1: The quantities a_k of an object B can be determined by counting how may times a free moving k-dimensional test element hits a k-dimensional object element. Let H_k be the number of these hits. Then we obtain

$$a_k = H_k / N_k^{(n)} .$$

Now, we consider two objects B and C given by two finite sets B_0 and C_0 of object points. The intersection $B_0 \cap C_0$ of these two sets is either empty or builds up a new object which we will call *intersection* of the two objects and denote by $B \cap C$.

Given an object B fixed in Z^n, and an object B' freely movable in Z^n. Our interest is to determine the sum

$$S_k(B,B') = \sum_{B'} a_k(B \cap B')$$

with respect to all translations and tiltings of B' for which the intersection with B is non-empty. This task seems to be unsolvable at the first glance.

But the problem is solvable because the numbers a_k of an object resulting from intersection can be determined by theorem 3.2.-1: After we have yielded a non-empty intersection $B \cap B'$, we assume this intersection as fixed in Z^n and determine $a_k(B \cap B')$ by counting the hits with a k-dimensional free movable test element e_k. That means to count how many times the equation

$$(B \cap B') \cap e_k = e_k = B' \cap (B \cap e_k)$$

is fulfilled.

Instead of determination of $B \cap B'$ at first, we determine all elementary objects $B \cap e_k$ which are equivalent to e_k. The number of these elementary objects is given by $a_k(B)$.

Now we have to test the freely movable object B' against all of these $a_k(B)$ fixed elementary objects. The result is the same as the result of testing the fixed object B' against $a_k(B)$ freely movable elementary objects. Therefore it follows the *intersection formula* [Vo87a,Vo88]

<u>Formula 3.2.-1:</u>

$$S_k(B,B') = N_k^{(n)} a_k(B)\, a_k(B') = 2^{n-1} k!\,(n-k)!\, a_k(B)\, a_k(B') .$$

This formula can be generalized to any linear combination of the a_k's. Let

$$R(B) = \sum_{k=0}^{n} r_k\, a_k(B)$$

be such an expression. Then we obtain the generalized intersection formula

Formula 3.2.-2:

$$S_R(B,B') = \sum_{k=0}^{n} r_k\, S_k(B,B') = \sum_{k=0}^{n} r_k\, N_k^{(n)}\, a_k(B)\, a_k(B')$$

$$= \sum_{k=0}^{n} r_k\, 2^{n-1}\, k!\, (n-k)!\, a_k(B)\, a_k(B')$$

as the main result of our counting theory. The theory given here is so to speak a "discrete integral geometry" in comparison to the integral geometry of E^n [Bl55, Sa76]. There the intersections between "figures" are investigated and measures like area A, perimeter P, or total curvature T of figures yielded by intersection are determined. For illustration, the three main formulas of two-dimensional integral geometry in E^2 shall be given here [Sa76]:

$$I_A = \int_{X'} A(X \cap X')\, dX' = 2\pi A A' ,$$

$$I_P = \int_{X'} P(X \cap X')\, dX' = 2\pi (A P' + P A') ,$$

$$I_T = \int_{X'} T(X \cap X')\, dX' = 2\pi (A T' + P P' + T A') .$$

The quantity $T(X)$ of a figure X is determined by $2\pi(N_o+N_i)$. Here, N_o is the number of outer boundaries (i.e. the number of components of X), and N_i is the number of inner boundaries of X (i.e. the number of holes).

3.2.3 Applications of Intersection Formula

As a first example, we will investigate the intersection of two objects in Z^2. The single objects B and B' must not be connected but we assume that they contain no holes. The Euler number of such an object is

$$\psi^{(2)} = \varepsilon - \varkappa + \mu = k$$

with μ as number of core meshes and k as number of components building up the object. For two-dimensional incidence structures $\Gamma(\nu,\lambda)$, the intersection formula gives the general formulas

$$S_\varepsilon = \nu\varepsilon\varepsilon' , \qquad S_\varkappa = 2\varkappa\varkappa' , \qquad S_\mu = \lambda\mu\mu'$$

because there are ν orientations for point-wise hitting, two orientations for edge-wise hitting, and λ orientations for mesh-wise hitting [Vo87a, Vo88]. For Z^2 as special case, it follows

$$S_\varepsilon(B,B') = 4\,\varepsilon\,\varepsilon' ,$$

$$S_\varkappa(B,B') = 2\,\varkappa\,\varkappa' ,$$

$$S_\mu(B,B') = 4\,\mu\,\mu'$$

for $B=B(\varepsilon,\varkappa,\mu)$ and $B'=B'(\varepsilon',\varkappa',\mu')$. Therefore, we obtain

$$S_\psi(B,B') = \sum_{B'} \psi(B \cap B') = 4\,\varepsilon\,\varepsilon' - 2\,\varkappa\,\varkappa' + 4\,\mu\,\mu' .$$

Because the intersections $B \cap B'$ also do not contain holes, $\psi(B \cap B')$ is the number of components of $B \cap B'$, and $S_\psi(B,B')$ yields the number of all components arising by intersections of the fixed object B and the free movable object B'. By this result, we can determine the mean values of any count measures, for instance

$$\overline{\varepsilon} = \frac{S_\varepsilon(B,B')}{S_\psi(B,B')} = \frac{4\,\varepsilon\,\varepsilon'}{4\,\varepsilon\,\varepsilon' - 2\,\varkappa\,\varkappa' + 4\,\mu\,\mu'} .$$

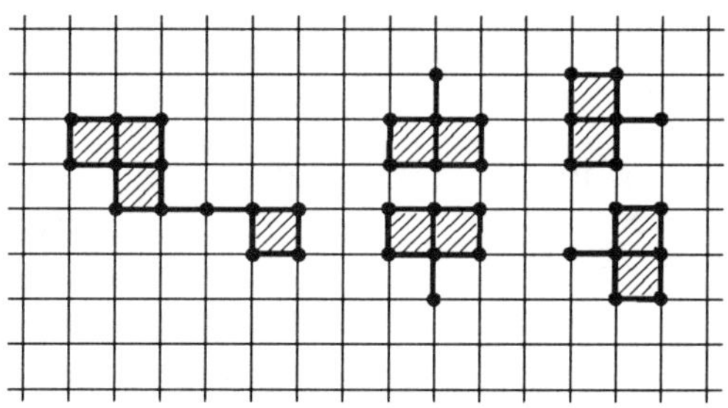

Fig. 3.2.-2: Fixed object $B=(\varepsilon,\varkappa,\mu)=(13,16,4)$ and some locations of the movable object $B'=(\varepsilon',\varkappa',\mu')=(7,8,2)$

Using figure 3.2.-2, it can be shown, that the intersection formula is very helpful to determine the sums S_ε etc. because the direct counting is cumbersome even by means of computers. We obtain for this example

$$S_\varepsilon = 4 \cdot 13 \cdot 7 = 364 ,$$

$$S_\varkappa = 2 \cdot 16 \cdot 8 = 256 ,$$

$$S_\mu = 4 \cdot 4 \cdot 2 = 32 ,$$

$$S_\psi = S_\varepsilon - S_\varkappa + S_\mu = 140 .$$

In section 2.4.4, we have found that $c_{12}=l=2\varkappa-4\mu$ is a linear measure for objects in Z^2 comparable to the perimeter of figures in E^2. Our main intersection formula yields

$$S_l = 2\,S_\varkappa - 4\,S_\mu = 4\,\varkappa\,\varkappa' - 16\,\mu\,\mu' \;.$$

Taking into account that $2\varkappa=l+4\mu$ and $2\varkappa'=l'+4\mu'$, we obtain

$$\begin{aligned} S_l &= l\,l' + 4\,\mu\,l' + 4\,l\,\mu' \\ &= 4(\mu\,l' + l\,\mu')\left[1 + \frac{l\,l'}{4\,\mu\,l' + 4\,l\,\mu'}\right] \;. \end{aligned}$$

For large objects (or for a small lattice distance used for digitization of two-dimensional Euclidean figures), the term $ll'/(4\mu l'+4l\mu')$ can be neglected. Then, the resulting formula $S_l=4(\mu l'+l\mu')$ is analogous to the above given formula $I_P=2\pi(AP'+PA')$ of integral geometry.

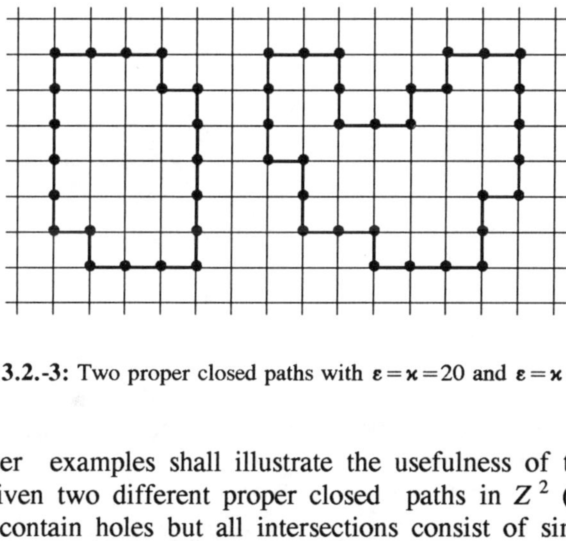

Fig. 3.2.-3: Two proper closed paths with $\varepsilon=\varkappa=20$ and $\varepsilon=\varkappa=30$

Two further examples shall illustrate the usefulness of the intersection formula. Be given two different proper closed paths in Z^2 (figure 3.2.-3). These objects contain holes but all intersections consist of simple connected objects without holes. Because $\mu=\mu'=0$, $\varepsilon=\varkappa$, $\varepsilon'=\varkappa'$, we obtain

$$S_\psi = S_\varepsilon - S_\varkappa = 2\,\varepsilon\,\varepsilon' \;.$$

The last equation follows because $\varepsilon=\varkappa$ for all proper closed paths. The mean values of the objects obtained by intersection are

$$\bar{\varepsilon} = \frac{S_\varepsilon}{S_\psi} = 2 \quad , \qquad \bar{\varkappa} = \frac{S_\varkappa}{S_\psi} = 1$$

independent of the shape and size of the given paths. The number S_ψ of all intersection objects only depends on the size of the two paths.

Finally, we investigate a square-like object with $\varepsilon = s^2$ points, $\varkappa = 2s(s-1)$ edges, and $\mu = (s-1)^2$ meshes in Z^2. This object can be located at pleasure within a square-like window with $\varepsilon' = s'^2$, $\varkappa' = 2s'(s'-1)$ edges, and $\mu' = (s'-1)^2$ meshes. It is possible that the object only lies partly in the window. How large are the intersected objects and which alteration arises by the window boundaries? It is

$$S_\varepsilon = 4\,\varepsilon\,\varepsilon' = 4\,s^2\,s'^2\ ,$$

$$S_\varkappa = 2\,\varkappa\,\varkappa' = 8\,s\,s'\,(s-1)\,(s'-1)\ ,$$

$$S_\mu = 4\,\mu\,\mu' = 4\,(s-1)^2\,(s'-1)^2\ .$$

Because

$$S_\psi = S_\varepsilon - S_\varkappa + S_\mu = 4\,(s+s'-1)^2\ ,$$

we obtain

$$\bar{\varepsilon} = \frac{S_\varepsilon}{S_\psi} = \frac{s^2\,s'^2}{(s+s'-1)^2}$$

so that only with $s' \to \infty$ the "right" value s^2 follows for the size of the square-like object.

3.2.4 Count Formulas

We have seen that some important topological and geometrical quantities are locally computable by means of counting of points, edges, meshes etc. Now, we will give an algorithmical approach to determine the a_k's for objects in Z^n.

For that, we shift an elementary n-cell (the *test cell* as for instance an edge in Z^1, a mesh in Z^2, a cell in Z^3) through the n-dimensional image box to be investigated (for Z^2 examples are given in figure 3.2-2/3). Each n-cell of the box is proved by the n-dimensional test cell, and we ask which contribution the single cell gives to the number of points, meshes etc.

In this investigation, we do not consider all tiltings of the n-dimensional test cell because all tiltings are identical. Further, we do not take into account the multiplicity of hits with k-dimensional object elements. In Z^2, we count only the number N_{config} of hits between object points and the upper left point of the test cell, and hits between object edges and the upper and left edge of the test cell (see figure 3.2.-4).

configu-ration	number	contribution to ϵ	κ	μ
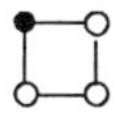	N_1	1	0	0
	N_2	1	1	0
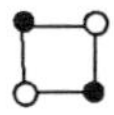	N_3	1	0	0
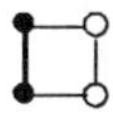	N_4	1	1	0
	N_5	1	1	0
	N_6	1	2	0
	N_7	1	1	0
	N_8	1	2	1

Fig. 3.2.-4: Elementary configurations in Z^2

If the upper left point of a single two-dimensional test cell is not an object point then we count neither points nor edges nor meshes in the corresponding configuration. Only eight configurations have to be investigated. Their contributions to the count results of $\varepsilon, \varkappa, \mu$ are listed in figure 3.2.-4. All values a_k and all linear combinations of a_k's can be determined by the values N_i where i goes from 1 to 2^j with $j=2^n-1$. We obtain for two-dimensional objects for instance the following expressions:

$$\begin{aligned}
\varepsilon &= N_1 + N_2 + N_3 + N_4 + N_5 + N_6 + N_7 + N_8 \,, \\
\varkappa &= N_2 + N_4 + N_5 + 2N_6 + N_7 + 2N_8 \,, \\
\mu &= N_8 \,, \\
\psi &= N_1 + N_3 - N_6 \,.
\end{aligned}$$

If we assume that the objects to be investigated contain no holes, the Euler number ψ is equal to the number of objects. In that case, we can determine the object number, and also the mean object size (i.e. the mean number of points per object), the sum $P=\varkappa-2\mu$ as total length of object border meshes, the mean perimeter $P/\psi=(N_2+N_4+N_5+2N_6+N_7)/(N_1+N_3-N_6)$ etc. without any object segmentation or contur detection (see also [Mi69, Gr71]).

One of the earliest count formula for $\chi=\varepsilon-\varkappa+\mu-\zeta$ in the three-dimensional case was given by Park and Rosenfeld:

> "A three-dimensional volume of 'digital space' may be represented by a stack of planes each of which contains a cross-section of the volume. Objects in such a volume can be specified by assigning the value 1 to their points, while points in the space between or within the objects have values 0... The genus can be computed by performing a local counting operation over the given volume. Let
>
> C_1 be the number of 1's
> C_2 be the number of 1 by 1 by 2 patterns of 1's (in all orientations)
> C_4 be the number of 1 by 2 by 2 patterns of 1's (in all orientations)
> C_8 be the number of 2 by 2 by 2 patterns of 1's.
>
> Then $E^{(1)}=C_1-C_2$, $E^{(2)}=C_1-C_2+C_4$, $E^{(3)}=C_1-C_2+C_4-C_8$...The proof is by induction on the number of 1's in the given volume." [Park and Rosenfeld, Pa71]

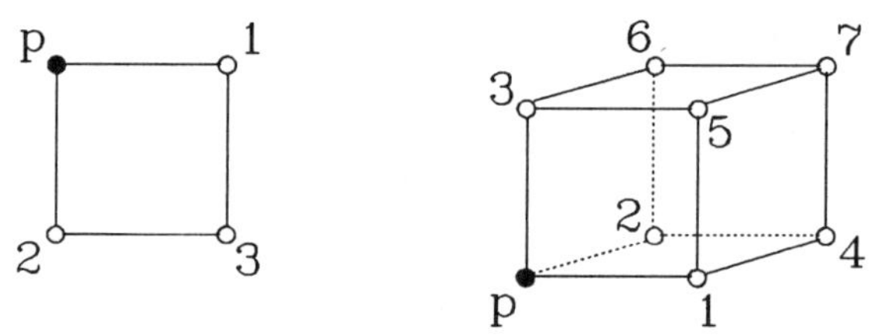

Fig. 3.2.-5: Point numbering for logical algorithms of ψ-determination

We can investigate the configurations to be counted also by another method which is typical for binary images. In figure 3.2.-5, a point numbering is given for two-dimensional configurations of figure 3.2.-4. Let the Boolean variable x_i be equal to 1 if the point i is an object point, and equal to 0 otherwise. The total number of object points in the binary image to be investigated is then

$$\varepsilon = \sum x_p .$$

An object edge is found if x_p and x_1 or x_2 are equal to one. Finally, to find an object mesh, all four points of the configuration must be object points, i.e. $x_p=x_1=x_2=x_3=1$. Therefore, we obtain

$$\varkappa = \sum (x_p x_1 + x_p x_2) \quad , \qquad \mu = \sum x_p x_1 x_2 x_3 \ .$$

This expression can be reformulated:

Formula 3.2.-3:

$$\begin{aligned} \psi^{(2)} &= \sum \left[x_p - x_p(x_1 + x_2) + x_p x_1 x_2 x_3 \right] \\ &= \sum_{x_p=1} \left[1 - x_1 - x_2 + x_1 x_2 x_3 \right] \\ &= \sum_{x_p=1} \left[\overline{x_1}\,\overline{x_2} - x_1 x_2 \overline{x_3} \right] \quad . \end{aligned}$$

Here, the symbols $\overline{x_i}=1-x_i$ are the logically negated values of x_i. The sign + (also the summation sign) and the minus sign mean arithmetic operations. This simple expression was already given by Lee and Rosenfeld in 1986 [Le86a]. The above derived formula $\psi=N_1+N_3-N_6$ yields the same result because

$$N_1 + N_3 - N_6 = \sum_{x_p=1} \left[\overline{x_1}\,\overline{x_2}\,\overline{x_3} + \overline{x_1}\,\overline{x_2}\,x_3 - x_1 x_2 \overline{x_3} \right] \quad .$$

In Z^3, we have to investigate 128 configurations so that the corresponding expressions will become more complicated. Lee and Rosenfeld have summarized existing algorithms to compute the Euler number of three-dimensional objects and have developed the following expression [Le86a]:

Formula 3.2.-4:

$$\psi^{(3)} = \sum_{x_p=1} \left[\overline{x_1}\,\overline{x_2}\,\overline{x_3} + (1 - \overline{x_1}\,\overline{x_2}\,\overline{x_3})(F - G) \right] \quad ,$$

$$F = (1 - x_7 x_4 x_5 x_6)(x_1 x_2 x_3 - x_1 x_2 \overline{x_4} - x_1 x_3 \overline{x_5} - x_2 x_3 \overline{x_6}) \quad ,$$

$$G = x_7 (1 - \overline{x_4}\,\overline{x_5}\,\overline{x_6})(x_1 x_2 \overline{x_3}\,\overline{x_4} + x_1 \overline{x_2}\,x_3 \overline{x_5} + \overline{x_1}\,x_2 x_3 \overline{x_6}) \quad .$$

3.2.5 Stochastic Images

The points of a random binary image (which we will call a *stochastic image*) build up black random objects in a white background or white random holes in a black background (figure 3.2.-6). We are interested in some topological features and count measures as functions of the only independent variable p of this model. The value of p is the probability that any picture point is black whereby the colors of different points shall be independent from one another.

Fig. 3.2.-6: Stochastic binary images with probability p=0.1, 0.2, 0.4, and 0.8

We assume that we have toroidally closed incidence structures Z^n with α points. The expectation value $E(\varepsilon)$ of the numbers of black points is then

$$E(\varepsilon) = \alpha p \ .$$

In two-dimensional incidence structures $\Gamma(\nu,\lambda)$ with α points, we have $\alpha\nu/2$ edges and $\alpha\nu/\lambda$ meshes. The probability that all of c anyhow selected points are black is p^c. Therefore, the expectation value for number of object edges in the image is

$$E(\varkappa) = \frac{\alpha\nu}{2} p^2 \ ,$$

and the expectation value for the number of object meshes containing λ object points is

$$E(\mu) = \frac{\alpha\nu}{\lambda} p^\lambda \ .$$

The expectation value of the Euler number of the image is then [Vo88, Vo89]

<u>Formula 3.2.-5:</u>

$$E(\psi^{(2)}) = E(\varepsilon - \varkappa + \mu) = E(\varepsilon) - E(\varkappa) + E(\mu) = \alpha\left(p - \frac{\nu}{2}p^2 + \frac{\nu}{\lambda}p^\lambda\right) \ .$$

Because of $1/\nu + 1/\lambda = 1/2$, the value of $E(\psi^{(2)})$ is equal to zero for $p=0$ and $p=1$. In the first case, we have a complete white image without objects. For $p=1$, we have a complete black image (toroidally closed), and the Euler number of such an image is equal to the Euler characteristic of the incidence structure themself, i.e. $\psi=\chi=0$. In table 3.2.-1, some values of $E(\psi^{(2)})/\alpha$ are given for the three structures $\Gamma(\nu,\lambda)$.

Table 3.2.-1: Expectation values $E(\psi^{(2)})/\alpha$ as function of probability p for $\Gamma(\nu,\lambda)$

p	$(\nu,\lambda)=(3,6)$	$(\nu,\lambda)=(4,4)$	$(\nu,\lambda)=(6,3)$
0	0	0	0
.1	.0850	.0801	.0720
.2	.1400	.1216	.0960
.3	.1653	.1281	.0840
.4	.1620	.1056	.0480
.5	.1328	.0625	0
.55	.1101	.0365	-.0248
.60	.0833	.0096	-.0480
.65	.0539	-.0165	-.0683
.70	.0238	-.0399	-.0840
.75	-.0048	-.0586	-.0938
.8	-.0289	-.0704	-.0960
.9	-.0493	-.0639	-.0720
1	0	0	0

The general character of these functions is similar in all three cases: Starting with $E(\psi^{(2)})=0$, the values arise continuously, i.e. more and more planar objects with outer boundaries exist. For greater probabilities, there are also objects with inner boundaries. Simultaneously, the objects merge and the number of outer boundaries decreases.

For a *critical probability* p^*, the expectation values are equal to zero. Then, for increasing probabilities, the numbers of inner boundaries are greater than the number of outer boundaries so that $E(\psi^{(2)})$ will become negative. But for $p=1$, the inner boundaries disappear also, and the expectation value is equal to zero once more. The critical probabilities are

$$p^*(3,6) = 0.741\,270\,910\ldots$$
$$p^*(4,4) = 0.618\,033\,988\ldots$$
$$p^*(6,3) = 0.5 \quad .$$

Because $E(\psi^{(2)})$ is equal to the expectation value of the difference between the number of outer and inner boundaries, i.e. equal to the expectation value of the total curvature of the image, the total curvature of all boundaries of the image disappears for $p=p^*$. That means that the boundaries do not know whether they shall turn in clockwise or counter-clockwise direction.

In physics, the model of stochastic distributed black and white points are used for investigation of *percolation* phenomena which are the reason for phase transitions in two-dimensional and three-dimensional lattices. For critical percolation probabilities p_p, an "infinite" black object suddenly arises which reaches all image borders. The theoretically determined or by simulation experiments estimated percolation probabilities are $p_p(3,6)\approx 0.6962$, $p_p(4,4)\approx 0.59275\pm 0.00003$, and $p_p(6,3)=0.5$ (see for instance [Se78, Es80, Ke82, St85, Bu91]).

There are close connections between physical percolation theory and some basical ideas of image processing as the following quotation shows.

> "In percolation theory, one studies random configurations of Z^2 in which each vertex can be in two possible states: it is either *open* or *closed*. An *open path* is a path whose elements are all open. Two open vertices z and z' are said to be *connected* if there exists an open path $(\pi_0,\ldots,\pi_n)$ such that $\pi_0=z$ and $\pi_n=z'$. An *open cluster* is a maximal set of connected open vertices... Typical questions in percolation theory are the following. (i) Does there exist an infinite open cluster? (ii) Is the origin (or any particular vertex) contained in an infinite cluster? (iii) How many infinite clusters arise?" [Meester and Nowicki, Me89]

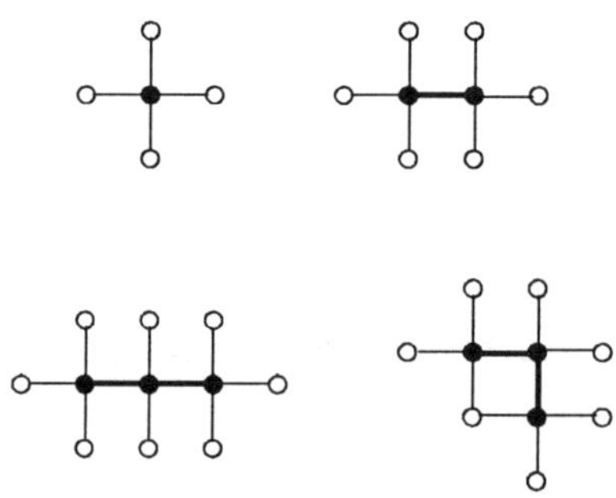

Fig. 3.2.-7: Small clusters and their surroundings

The probability for small clusters can be computed explicitly whereby all possible orientations must be taken into account (figure 3.2.-7). It follows

$$E(n_1) = \alpha p (1-p)^4 ,$$

$$E(n_2) = 2\alpha p^2 (1-p)^6 ,$$

$$E(n_3) = 2\alpha p^3 (1-p)^8 + 4\alpha p^3 (1-p)^7 .$$

In [Kl85a], some results are given about the exact number of different non-fitting connected objects of given point number ε (i.e. "small clusters") in two-dimensional lattices. Thereby "non-fitting" means that two objects cannot be fitted by any translations, rotations, and reflections (table 3.2.-2).

Table 3.2.-2: Number of different two-dimensional objects

ε	$(\nu,\lambda)=(4,4)$	$(\nu,\lambda)=(6,3)$	$(\nu,\lambda)=(3,6)$
1	1	1	1
2	1	1	1
3	2	3	1
4	5	7	3
5	12	22	4
6	35		12
7			24

In the two-dimensional case, all expressions up to $E(n_{17})$ are known [St79]. By means of computer aided simulation experiments in a 4000·4000 triangular lattice $\Gamma(6,3)$, the approximation

$$n_s(p_p) = n_s\left(\frac{1}{2}\right) \approx const \cdot s^{-2}$$

was found. This result points out that such random images are characterized by fractal properties [St79]. Ahuja has given approximative numerical expressions for the expected number $E(N)$ of connected black objects in different lattices and mosaics [Ah78, Ah78a]. However in contrast to the formula for $E(\psi^{(2)})$, there is no exact expression for $E(N)$.

The investigation of Euler number can be performed also for higher dimensional incidence structures (for percolations see [Gr89]). In a toroidally closed Z^3 with α points, we have $\varkappa=3\alpha$ edges, $\mu=3\alpha$ meshes, and $\zeta=\alpha$ cells. Therefore, the expectation values corresponding to the black point probability p are

$$\begin{aligned} E(\varepsilon) &= \alpha p \quad , \\ E(\varkappa) &= 3\alpha p^2 \quad , \\ E(\mu) &= 3\alpha p^4 \quad , \\ E(\zeta) &= \alpha p^8 \quad . \end{aligned}$$

We obtain

$$E(\psi^{(3)}) = \alpha(p - 3p^2 + 3p^4 - p^8)$$

as function which disappears for $p=0$ and $p=1$. But we have also two other probabilities $p_1^* \approx 0.394$ and $p_2^* \approx 0.886$ where $E(\psi^{(3)})$ disappears. Possible connections to the three-dimensional percolation theory are only weak because the percolation probability for the three-dimensional lattice is $p_p \approx 0.3116$ [Эф82, Ca85, Bu91].

Finally, we can write down the formulas for expectation values in randomly occupied n-dimensional images. Because of the matching formula 2.1.-1, we obtain

$$a_k^{(n)} = a_0 \frac{b_{0k}^{(n)}}{b_{k0}^{(n)}} = a_0 \binom{n}{k} .$$

It follows that

$$E(a_k^{(n)}) = \alpha \binom{n}{k} p^{2^k}$$

with $\alpha = a_0$, and therefore the expectation value

<u>Formula 3.2.-6:</u>

$$E(\psi^{(n)}) = \alpha \sum_{k=0}^{n} (-1)^k \binom{n}{k} p^{2^k}$$

of the Euler number is given as function of the probability p. This function is equal to zero for $p=0$ and $p=1$, independent from dimension n. An approximative solution for the first non-trivial critical probability in Z^n can be obtained from $p(1-np+n(n-1)p^3/2-+\dots)=0$ or from

$$p_1^*(n) \approx \frac{1}{n} + \frac{n-1}{2n^3} \quad .$$

3.3 Topology Preserving Operations

3.3.1 Topological Equivalence

Topology is that mathematical discipline which investigates the topological structure of spaces, point sets in such spaces, and properties which are invariant under topological mappings. In continuous spaces, a topological mapping is a one-to-one mapping between two topological spaces. For instance, a circle C and a square S are topologically equivalent in E^2.

In incidence structures, especially in Z^n, we have to define topological equivalence with respect to the discrete nature of the structures. At first, there are only such trivial one-to-one neighborhood preserving mappings in incidence structures as described in section 3.2.1 by the notion of tilting. Therefore one could conclude that there are no topological equivalences in Z^n.

In this sense, topological equivalence in discrete structures cannot be defined by means of any mappings. Instead of that, we have to investigate in which circumstances two objects B and B' are topologically equivalent to one another.

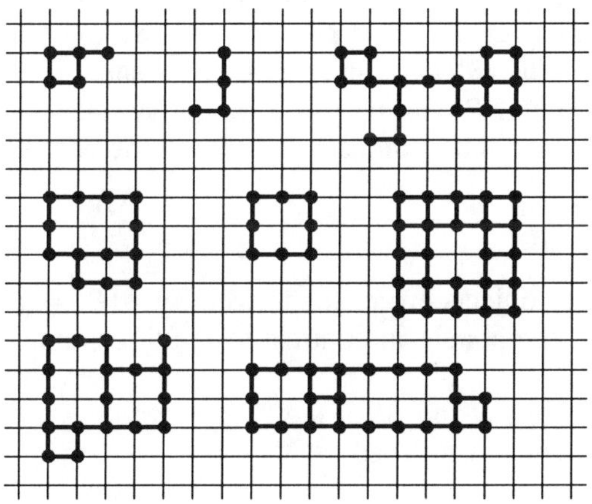

Fig. 3.3.-1: The objects in the first, second, third row are topologically equivalent to one another

In figure 3.3.-1, the two-dimensional connected objects of each row are assumed as equivalent to one another. But neither the number of points nor the number of edges or meshes are sufficient to characterize such equivalences. The only quantity which is the same for each row is the Euler number $\psi=\varepsilon-\varkappa+\mu$ ($\psi=1$ in the first row, $\psi=0$ in the second row, $\psi=-1$ in the third row).

Nevertheless, we cannot say that $\psi(B)=\psi(B')$ is sufficient to guarantee the equivalence between B and B'. For instance, a composed object $B=B_1 \cup B_3$ with B_1 as object of the first row, B_3 as object of the third row, and $B_1 \cap B_3 = \varnothing$ has the same Euler number $\psi(B)= \psi(B_1)+\psi(B_3)=0$ as any object of the second row. Therefore, we require that B and B' consist of the same number of components and have the same Euler number to be equivalent. But this is also not sufficient as figure 3.3.-2 shows.

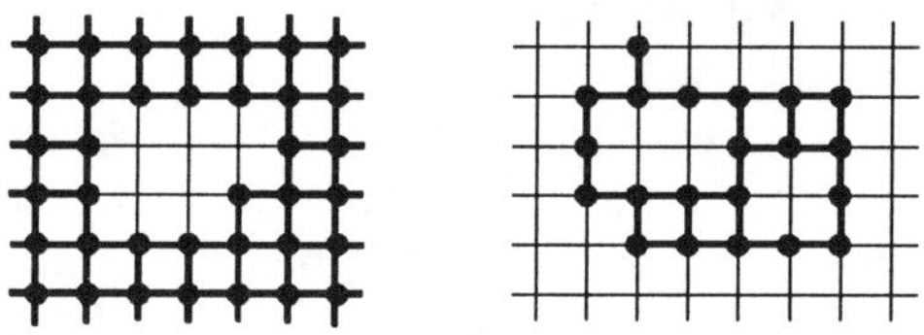

Fig. 3.3.-2: Two connected objects with the same Euler number

The left object B is a connected toroidal object (in a finite two-dimensional neighborhood structure) with only an inner border mesh. Therefore $\psi(B)=-1$. The right connected object B' is also characterized by $\psi(B')=-1$ because it has one outer border mesh and two inner border meshes. The only topological difference between B and B' can be found when we take into account also the topology of the complementary objects defined by the complementary sets

$$B_c = \overline{B} \qquad , \qquad B_c' = \overline{B'} \quad .$$

These complementary objects are characterized by $\psi(B_c)=\psi(B_c')=+1$, but B_c consists only of one component in contrast to the three components of B_c'.

Def. 3.3.-1: Two objects B and B' of the Z^2 are *topologically equivalent* if the following conditions are fulfilled:

- B consists of the same set $\{B_1, B_2, \ldots, C_1, C_2, \ldots\}$ of (connected) components and complementary components (in short "con-components") as B' which is composed of con-components $B_1', B_2', \ldots, C_1', C_2', \ldots$

- There is a one-to-one attachment between the con-components of B and the con-components of B' so that two attached con-components are equivalent to one another.

Def. 3.3.-2: Two con-components B and B' are *topologically equivalent* if the following conditions are fulfilled:

- It is $\psi(B)=\psi(B')$.
- The complementary objects B_c and B_c', both defined with respect to B and B', respectively, are topologically equivalent.

These two definitions are recursive in nature. But the recursion breaks off after the second step because the complement of a complementary object is always connected (see section 1.3.4, theorem 1.3.-1).

We must take into account that these definitions are not sufficient to describe "total topological equivalence": A proper simple path of object points in Z^3 is locally equivalent to a knotted path with respect to definitions 3.3.-1/2, but there is no global topological equivalence between these two paths.

3.3.2 Simple Points

In this section, we will investigate possibilities of alterations of objects in such a manner that the original object and the new object are topologically equivalent to one another (also in respect to a global view point). Because we cannot give any functional description of such alterations, we have to search for algorithms which work in n-dimensional incidence structures.

> "Let P be a point of S, and let $N(P)$ denote the set of eight neighbors of P ... A point of S that has the properties ... $S\cap N(P)$ has the same number of components as $S\cap[N(P)\cup\{P\}]$ ... will be called *simple*." [Rosenfeld, Ro77]

> "In 2D we call a point $p \in S$ *simple* if (the number O of objects) $O([S\cap N(p)]\cup\{p\})=O(S\cap N(p))$ and $O([\ \overline{S}\ \cap N(p)]\cup\{p\})=O(\ \overline{S}\ \cap N(p))$ ... In 3D we require additionally that (the number H of holes) $H([S\cap N(p)]\cup\{p\})=H(S\cap N(p))$ and $H([\ \overline{S}\ \cap N(p)]\cup\{p\})=H(\ \overline{S}\ \cap N(p))$ ... Deletion of a simple point from a 2D binary image is a topology preserving operation." [Morgenthaler, Mo81]

> "Digital topology is the study of the topological properties of binary image arrays. The subject has applications in the theory of image thinning, border following and object counting...A black point p in a two-dimensional digital picture is called a *simple* point if its deletion does not alter the topology of the picture." [Kong and Rosenfeld, Ko86a]

In many papers, it is contemplated that an image consists of black and white points. In our theory, we will investigate point sets B which build up objects of an incidence structure. Of course, we can denote the object points $p \in B$ as black points and the points $q \notin B$ as white points, but this is only another naming. Rosenfeld has pointed out that a point is *simple* if and only if its color (black or white) does not influence the number of foreground and background components [Ro79, Ec88]. In this sense, we will give a definition of simple points.

Def. 3.3.-3: A *simple object point* $p \in B$ of an object B of an incidence structure Z^n is defined by the condition that the two objects B and $B' = B - \{p\}$ are topologically equivalent to one another.

configuration	$\Delta\psi$	$\Delta\psi_c$
○ ○ ● ● ● ● ● ● ○	1	0
○ ○ ● ○ ● ● ● ● ●	0	0
○ ○ ○ ○ ● ● ○ ○ ○	0	0
○ ● ○ ● ● ● ○ ● ○	3	1
● ○ ● ○ ● ○ ● ● ●	0	−2

Fig. 3.3.-3: Alterations of Euler numbers $\Delta\psi$ and $\Delta\psi_c$

Because the alteration of the connectivity of an object and the alteration of the Euler number of an object are locally computable (both properties are to be investigated in respect of the foregoing definitions), we have to prove the simplicity of an object point in its local surrounding. The local computability follows from the linearity of the Euler number as sum over object elements.

With other words: if we compute the Euler number ψ_p of the 3^n-surrounding of the point $p \in B$ before and after the deletion of p, then the difference $\Delta\psi$ must be equal to zero.

In figure 3.3.-3, the alterations $\Delta\psi$ of the object B and $\Delta\psi_c$ of the complementary object are given for some 3·3-surroundings S_3 of the central point which has to be deleted. There are maximally 9 object points, 12 object edges, and 4 object meshes in S_3. Therefore, we obtain $\psi=\varepsilon-\varkappa+\mu$ and $\psi_c=\varepsilon_c-\varkappa_c+\mu_c$. It is to be seen that only the second and third configurations contain a simple central point because in these case we obtain $\Delta\psi=\Delta\psi_c=0$.

In [Ko86a], many references are given for different algorithms and methods to determine the simplicity of a point. But it is not the place here to review those methods because they are all variations in respect of the given definitions.

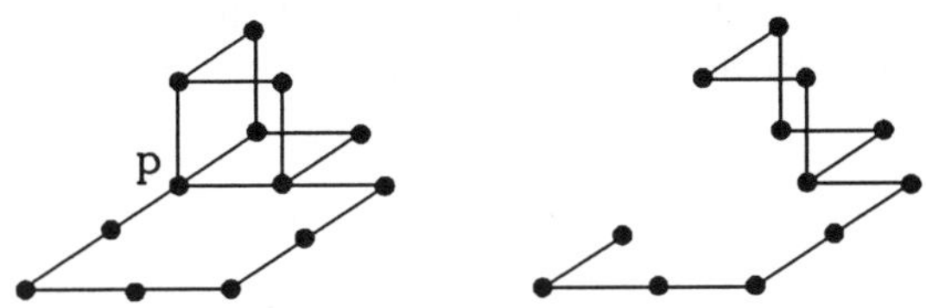

Fig. 3.3.-4: Is the point p a simple point of the 3-dimensional object?

The decision whether a point is simple or not is not so easy to obtain for three-dimensional objects as for two-dimensional objects. For example, in figure 3.3.-4 the point p shall be investigated. We see that both objects B and B' are connected, both objects contain only $S=1$ surface and exactly $T=1$ tunnel. The formula $\psi=\varepsilon-\varkappa+\mu-\zeta$ as well as the formula $\psi=S-T$ give for both objects the value $\psi=0$. Therefore, we can conclude that both objects are topologically equivalent. But Morgenthaler describes this example as follows:

> "For 3D binary images, the problem of defining topology preserving operations is more difficult since in addition to objects and cavities, there may also be holes. Certainly, we must require of a topology preserving operation that it not change the number of objects, cavities, holes, or the adjacency and surroundednes relations among objects and cavities...though, we can see that this is not enough...the hole in S is *different* than the hole in $S-\{p\}$." [Morgenthaler, Mo81]

It is not necessary to require that two topologically equivalent objects B and B' contain the same number of "identical, non-different" holes (i.e. tunnels) or cavities. If we accept that both objects in figure 3.3.-4 are topologically equivalent then the deletion of p is a topology preserving operation, and p is a simple point.

<u>Theorem 3.3.-1:</u> Let $p\epsilon B$ be a point of object B in the n-dimensional incidence structure Z^n, and let S_n be the 3^n-surrounding of the point $p\epsilon B$. We compute the differences

$$\Delta\psi = \psi\left(S_n \cap (B - \{p\})\right) - \psi\left(S_n \cap B\right)$$

$$\Delta\psi_c = \psi\left(S_n \cap \overline{B - \{p\}}\right) - \psi\left(S_n \cap \overline{B}\right)$$

of Euler number before and after deleting of p. If both differences $\Delta\psi$ and $\Delta\psi_c$ are equal to zero then p is a *simple point*, and the deletion of p is a *topology preserving operation.*

We obtain for the example of figure 3.3.-4 the following results before and after deleting of point p in the "upper half" of the 3^3-surrounding of p (see figure 3.3.-5):

Before:	$\varepsilon = 8$	$\varkappa = 10$	$\mu = 3$	$\zeta = 0$
	$\varepsilon_c = 19$	$\varkappa_c = 26$	$\mu_c = 8$	$\zeta_c = 0$
After:	$\varepsilon = 7$	$\varkappa = 6$	$\mu = 0$	$\zeta = 0$
	$\varepsilon_c = 20$	$\varkappa_c = 28$	$\mu_c = 9$	$\zeta_c = 0$

It is $\Delta\psi = \Delta\psi_c = 0$, and p is a simple point if all black points in the right side of figure 3.3.-5 are connected also after deletion of p.

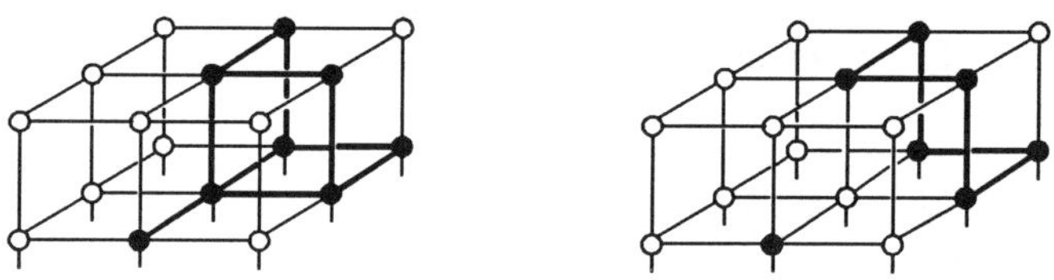

Fig. 3.3.-5: Upper half of the 3^3-surrounding S_p of figure 3.3.-4

A foregoing characterization of simple points for any finite dimensional Euclidean grid-point space was given by Tourlakis and Mylopoulos [To73], and Morgenthaler has investigated simple points in Z^2 and Z^3 [Mo81]. Three-dimensional skeletonization algorithms were investigated also in [Ar72, Lo80, Ts81, To82, Ge91].

3.3.3 Thinning

Thinning as understood in the common sense means the erosion of an object without topological alterations. Thereby one wishes that some properties of *thinning* or *skeletonization* are guaranted.

> "An ein Skelett werden im allgemeinen die folgenden Anforderungen gestellt:
> a) Das Skelett muß aus Linien der Breite eines Bildpunktes bestehen.
> b) Die topologischen Zusammenhänge des Skeletts müssen diejenigen des Originalbildes wiederspiegeln, d.h. die Anzahl der nach der festgelegten Metrik zusammenhängenden Komponenten und die Euler-Nummer müssen übereinstimmen.
> c) Die Skelettlinien müssen etwa in der Mitte der Objekte verlaufen.
> d) Bei dicken oder zerklüfteten Linien dürfen möglichst wenige kurze Verästelungen des Skeletts entstehen...
> e) Das Ergebnis der Skelettierung darf sich nach einer bestimmten Anzahl Iterationen des Skelettierungsalgorithmus nicht mehr ändern." [Zamperoni, Za89] [T5)]

It is clear that such requirements (see also [Da81, Ch87]) are dependent on practical intentions and do not reflect any theoretical considerations. For instance, it is irrelevant to our theory that the skeleton lies nearly in the "middle of the object", but it is important that the skeleton is topologically equivalent to the original object.

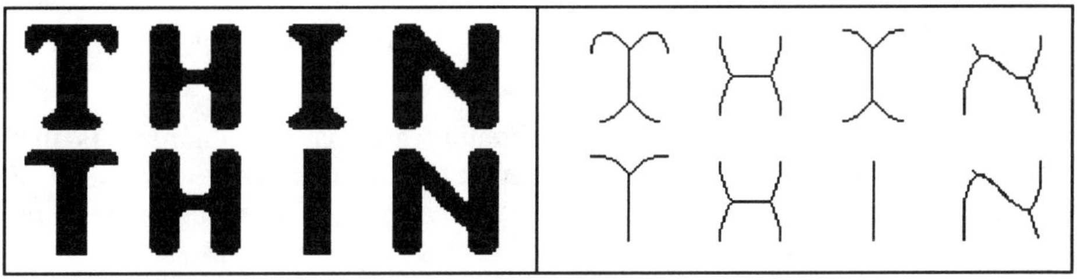

Fig. 3.3.-6: Original binary image (left) and skeletonization (right)

Skeletonization is often realized by sequential deleting of simple points. An essential practical point of view is that the end points of skeleton branches shall be preserved although they are simple points. Therefore, we have a difference between thinning and *shrinking* in that sense that an arc is invariant under thinning, but can be shrunk to a single point. On the other hand, closed curves are invariant under shrinking because the hole must be preserved.

"For a rigorous mathematical analysis of the sets resulting from thinning it would be highly desirable to obtain a well-defined result from the thinning process. Therefore parallel approaches are preferred not only in the interest of computational efficiency alone. Parallel removal of points from a set means that these points can be removed independently without influencing each other. Any well-defined method for parallel thinning should consider as candidates for removal only points which can be removed unconditionally." [Eckardt and Maderlechner, Ec88]

Eckardt uses the notion of *perfect points* to realize these requirements [Ec87]. Each perfect point has to be also a simple point. We define perfect points as follows:

Def. 3.3.-4: Let $p \epsilon B$ be a simple point of object B, and let S_p be the 3^n-surrounding of p. If $q \epsilon B$ is any other simple point in S_p, the point p is only then a *perfect* point if the deletion of q does not destroy the simplicity of p.

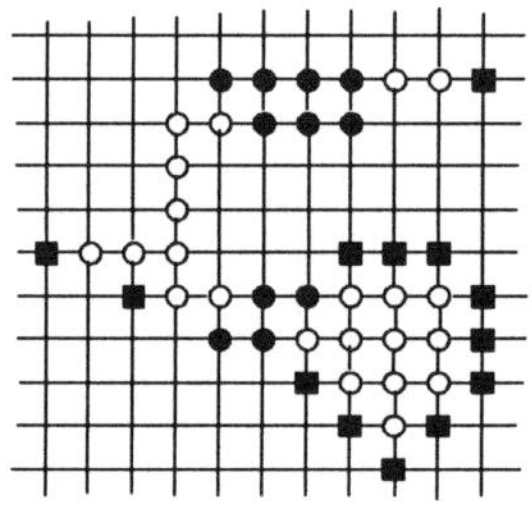

Fig. 3.3.-7: Object points (○,●,■), simple points (●,■), and perfect points (■)

The difference between sequential thinning and parallel thinning is that sequential thinning deletes one simple point at a time whereby the simplicity of a point depends on preceding deleted points, while parallel thinning operates on all perfect points simultaneously.

It is shown in figure 3.3.-7 that all perfect points are simple points and that they can be deleted without altering the topological structure of the object. Clearly, also this simultaneous deleting has to be carried out successively because after the first parallel deleting we obtain a new object, and this new object contains new simple points and perfect points.

The theoretical advantage consists in the fact that the new object yielded by deleting perfect points is unambiguously defined. The disadvantage is that the result of some thinning steps can obtain "skeleton parts of wide 2" because the simple points of such parts cannot be deleted simultaneously.

4 Geometrical Laws and Properties

4.1 Discrete Geometry

4.1.1 Geometry and Number Theory

At the end of the last century, the book "Geometrie der Zahlen" was published by H.Minkowski in Leipzig [Mi96]. Minkowski was mainly interested in number theory. But he had also a deep understanding of geometrical problems. The combination of number theoretical ideas and geometrical conceptions has been proven as a very fruitful approach to a new mathematical discipline namely to the *geometry of numbers* [Ca59].

The reason to devote our interest to Minkowski's work is that the Z^n is characterized by geometrical properties (distance of two points given by a metric) and by number theoretical properties (because all points have integer coordinate values).

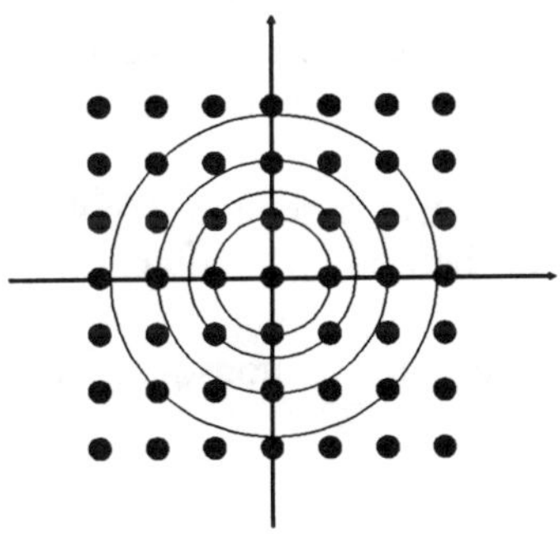

Fig. 4.1.-1: Lattice points and circles with $r^2=1,2,4,8$

One of the oldest problems in this field was already investigated by C.F.Gauß [Wi29]. He has solved the question: how large is the number of lattice points within a circle of radius r around the origin? This number $N(r)$ is defined by

$$N(r) = \sum_{m^2+n^2 \leq r^2} 1 .$$

We count for instance in figure 4.1.-1 the following numbers:

r^2	1	2	3	4	5	6	7	8	9
$N(r)$	5	9	9	13	21	21	21	25	29

Gauß has proven that the number $N(r)$ can be estimated by

$$N(r) = \pi r^2 + O(r) \ .$$

That means that $N(r)$ nearly describes the area of the circle $x^2+y^2=r^2$ up to a remaining term of order r. The further investigation of this remaining term is the content of the so-called *circle problem* of number theory . In 1962, it could be proven that [Kr81]

$$N(r) = \pi r^2 + O(r^\alpha) \qquad \text{with} \quad \alpha > \frac{24}{37} \ .$$

One presumes that the result $\alpha \geq \frac{1}{2} + \delta$ with very small δ will be optimal [Kr811].

It is clear that number theoretical questions (How many possibilities for number combinations with $a^2+b^2 \leq c^2$ for given c?) and geometrical questions (How many lattice points (x,y) within the circle with radius r?) are connected.

Because we deal in digital image processing mainly with integer valued point coordinates (in two and more dimensions), number theoretical laws will play an essential role in such investigations. For instance, the combinatorical geometry [Ha64] offers a few results which may be relevant to discrete topology and geometry as applied in image processing. We quote here three theorems as example for the two-dimensional orthogonal lattice:

- If k lattice points should form a regular k-polygon, then $k=4$ is the only solution. It follows, that there is no regular triangle in Z^2.

- If four lattice points determine a nonsquare rhombus with angle α, the quotient α/π is irrational.

- If α is an acute angle of a Pythagorean triangle of lattice points, the number α/π is irrational.

Similar problems as the circle problem above mentioned arise in Z^n where the number of lattice points in an n-dimensional sphere is of interest. For $n=3$, it is

$$N^{(3)}(r) = \frac{4}{3}\pi r^3 + O(r^\alpha) \qquad \text{with} \quad \alpha > \frac{22}{16} \ .$$

The general case in Z^n can be described by

$$N^{(n)}(r) = \frac{\pi^{n/2}}{\Gamma\left(\frac{n}{2}+1\right)} r^n + O(r^{n-1}) \ .$$

One of the main results in lattice point theory is Minkowski's *lattice point theorem*:

Theorem 4.1.-1: Each convex central symmetric figure of the Euclidian space E^n with its centre in origin and a volume $V \geq 2^n$ contains out of the origin at least two further lattice points (possible at the boundary).

A generalization of Minkowski's theorem was found by Blichfeldt with

$$2 \cdot ent\left[\frac{V(C)}{2^n}\right] + 1 \leq \varepsilon(C)$$

for the number $\varepsilon(C)$ of lattice points of the Z^n which are covered by a central symmetric convex Euclidean set C with volume $V(C)$ [Bl21, Wi90]. Neglecting the assumption of central symmetry, it was shown that [Bo72]

$$\varepsilon(C) \geq V(C) - \frac{1}{2} S(C)$$

where $S(C)$ is the content of the (n-1)-dimensional surface of the convex set C.

There is also an upper limit for $\varepsilon(C)$. For convex sets C in E^n, the inequality

$$\varepsilon(C) \leq \sum_{i=0}^{n} \sum_{j=1}^{\binom{n}{i}} V_i\left(C_j^i\right)$$

is valid, and there is equality for lattice boxes with edges parallel to the coordinate axis [Da51]. Here the $V_i(C_j^i)$ mean the i-dimensional volume of the projection of set C at one of the $\binom{n}{i}$ i-dimensional subspace $E_j^i \subset E^n$ with $j \in \{1 \ldots \binom{n}{i}\}$.

Another estimation for ε yields the wide elaborated theory of lattice polytopes [Mc77, Be85, Be85a, Gr87, Wi90]. The convex hull of a finite number of lattice points in Z^n is denoted as *lattice polytop*. The oldest results are Pick's theorem (see section 2.3.3, [Pi99]) and Blichfeldt's theorem [Bl21, Wi90]

$$\varepsilon(P) \leq n!\, V(P) + n$$

where P is a lattice polytop, $V(P)$ the volume of P, and $\varepsilon(P)$ the number of lattice points covered by P.

Unfortunately, the theory of lattice polytopes is productive only for convex sets, and also Minkowski's theorem does not have an immediate utility in image processing. We can only hope that the number theoretical ideas of lattice point theory will become of more importance in future.

An instructive result was given by Blichfeldt for any-shaped sets in the two-dimensional case [Bl14, Ca59, Bo71]:

Theorem 4.1.-2: Let F be a figure in the two-dimensional Euclidean space with area $A > N$ whereby N is an integer number. It is possible by translation that the translated figure F' covers at least $N+1$ lattice points.

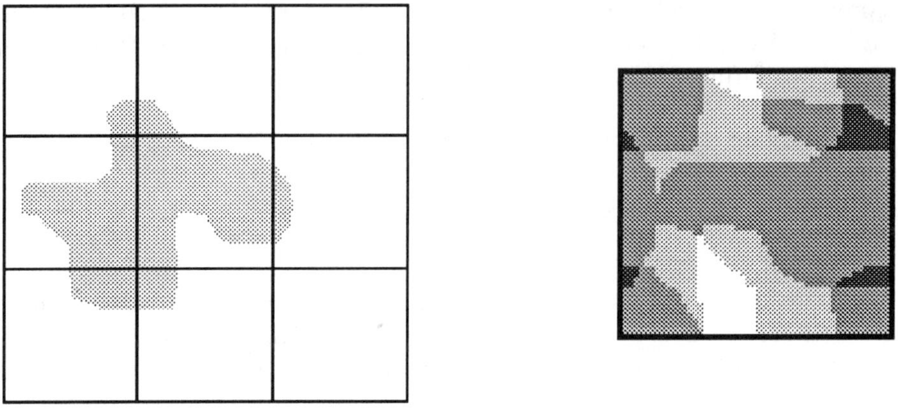

Fig. 4.1.-2: Example for the proof of Blichfeldt's theorem

The proof of *Blichfeldt's theorem* is relatively simple (see figure 4.1.-2): We cut the plane to small $1 \cdot 1$-squares so that there is exactly one lattice point in each square. Then we stack the squares one upon another in a single stack square S.

Now, an integer valued function $f(x,y)$ is defined for each point of S which will describe how often the point (x,y) will cut the given figure F in the corresponding squares. Therefore, the area of F can be obtained by

$$A = \int_{x=0}^{1} \int_{y=0}^{1} f(x,y)\; dy dx = \overline{f}$$

so that A corresponds to the mean value of the function f in respect of the square S. Because of $A > N$, this mean value is larger than N, and therefore it must be points (x,y) with $f(x,y) > N$. Taking into account that $f(x,y)$ is an integer function, it follows that at least one point (x',y') exists with $f(x',y') \geq 1$. Using the basic idea of this proof, the theorem can be generalized also to higher dimensions.

4.1.2 Minkowski Geometry

We will here shortly describe Minkowski's geometry [Bo34, Ro71, Le80]. In the set M^n of all real valued tupels $(x_1,\ldots,x_n)$, a *distance measure* $m(x_1,\ldots,x_n)$ may be given with the following properties:

$$m(x_1,\ldots,x_n) > 0 \quad \text{for} \quad (x_1,\ldots,x_n) \neq (0,\ldots,0) \;,$$

$$m(0,\ldots,0) = 0 \;,$$

$$m(tx_1,\ldots,tx_n) = |t|\, f(x_1,\ldots,x_n) \;,$$

$$m(x_1+y_1,\ldots,x_n+y_n) \leq m(x_1,\ldots,x_n) + m(y_1,\ldots,y_n) \;.$$

The points $(x_1,\ldots,x_n)$ which fulfill the inequality $m(x_1,\ldots,x_n)\leq c$ represent a central symmetrical convex set C in M^n where the origin $(0,\ldots,0)$ lies inside of C. For $c=1$, we denote C as *unit set* C_1. The distance between two points is defined by

$$d\big((x_1,\ldots,x_n) , (y_1,\ldots,y_n) \big) = m(x_1-y_1,\ldots,x_n-y_n)$$

and will be named by the notion *ray distance*. The ray distance defined by the distance measure m or by the unit set C_1 fulfills the conditions of a metric space given in definition 1.4.-1 so that M^n is a metric space - the so-called n-dimensional *Minkowski space*. The ray distance between two points $x,x+a\epsilon M^n$ is invariant in respect of translations, i.e.

$$d\big((x_1,\ldots,x_n) , (x_1+a_1,\ldots,x_n+a_n)\big) = m(a_1,\ldots,a_n) \;.$$

Therefore, translations are allowed motions in M^n. But any rotations in M^n defined by

$$\begin{pmatrix} x_1' \\ x_2' \\ \vdots \\ x_n' \end{pmatrix} = \begin{pmatrix} r_{11} & r_{12} & \cdots & r_{1n} \\ r_{21} & r_{22} & \cdots & r_{2n} \\ \vdots & \vdots & \vdots & \vdots \\ r_{n1} & r_{n2} & \cdots & r_{nn} \end{pmatrix} \begin{pmatrix} x_1 \\ x_2 \\ \vdots \\ x_n \end{pmatrix}$$

with an orthogonal matrix $R=(r_{kl})$ are only then possible motions if the symmetry of the unit set C_1 is taken into account.

In figure 4.1.-3, two unit sets are shown. One can seen that rotations only are possible by $\pm 180^0$ in the left example and $\pm 90^0$ in the right, respectively. Thereby we require that all distances $d(a,b)$, $d(o,a)$, $d(o,b)$ etc. remain constant. Solely reflections at symmetry axis of the unit sets are yet allowed.

We see that the non-Euclidian character of Minkowski's geometry yields the rigorous restriction of rotation possibilities which causes the anisotropy of the Minkowski space. If we require that the graph distances (see section 1.2.4) in Z^n remain constant by "rotations", then we have to guarantee the invariance of all distances shown in figure 4.1.-4. In fact, we have always the relationship

$$d(cx_1,\ldots,cx_n) = |c|\, d(x_1,\ldots,x_n)$$

for the distances between the origin $(0,0,\ldots,0)$ as rotation centre and the point $x=(x_1,\ldots,x_n)$ on the hand and between the origin and cx on the other hand for any integer c independent of the used graph metric in Z^n.

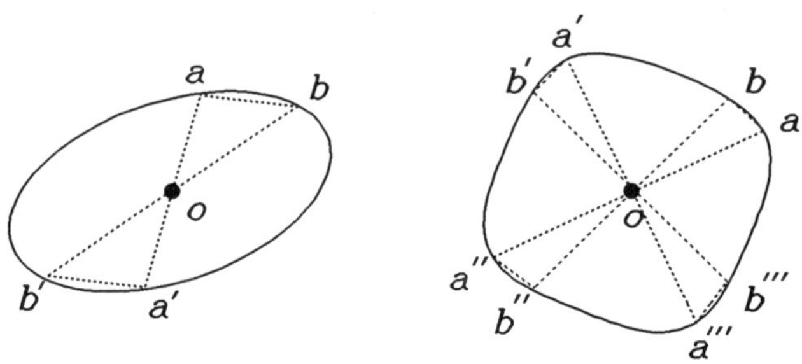

Fig. 4.1.-3: Unit sets in Minkowski plane with two-fold and four-fold rotation symmetry

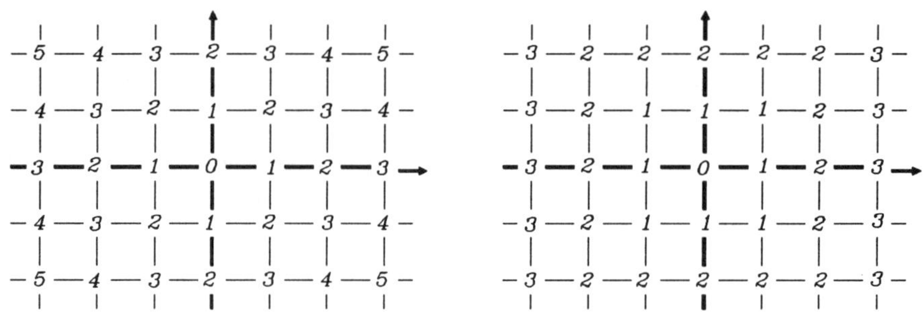

Fig. 4.1.-4: Graph distances defining Minkowski metrics

It is not a pecularity of the lattices used in image processing that only tiltings are allowed. Also in continuous point sets with an anisotropic Minkowski metric that property appears. Additionally, the requirement of one-to-one mappings for motions in lattices forces the restriction to tiltings also when we use the Euclidian metric.

All geometrical properties of a Minkowski space are complete determined by the unit set C_1 or by the distance measure $m(x_1,\ldots,x_n)$. Especially, the number π_M of Minkowski's geometry is defined as usually by $U_1=2\pi_M$ with U_1 as the perimeter of the unit set. It can be shown [Ro71, Le80] that

$$3 \leq \pi_M \leq 4$$

whereby $\pi_M=3$ is fulfilled for a regular 6-polygon used as C_1, and $\pi_M=4$ is fulfilled for a square.

It is an inadequate approach to use only the Euclidian geometry for lattices used in image processing. More appropriately, we have to develop consequently a *lattice geometry* and to study afterwards the relationships between the Euclidean space E^n and the orthogonal lattice Z^n. One can argue that as the underlying geometric space in image processing is discrete, there is a need to convert geometric concepts and algorithms from the continuous to the discrete domain. Such an approach has been done by A.Hübler [Hü87, Hü87a]:

> "Die diskrete Natur digitaler Bilder macht es erforderlich, angepaßte Begriffssysteme und Methoden zu entwickeln, denn eine direkte, unvermittelte Benutzung klassischer geometrischer Theorien ist nicht möglich bzw. den Erfordernissen nicht angepaßt...Die klare, auf F.Klein [Kl72] zurückgehende Konzeption, Geometrie als eine Invariantentheorie einer gegebenen Grundmenge bezüglich einer Transformationsgruppe zu begreifen, ist hier nicht das bestimmende Element ...Wir folgen der Auffassung von Böhm und Hertel [Bö87], die die *diskrete Geometrie* im Sinne von Klein als Invariantentheorie über einer Grundmenge mit einem System von *Figuren* bezüglich einer Transformationsgruppe verstehen." [Hübler, Hü89] T6)

4.1.3 Translative Neighborhood Structures

In this section, we describe the approach of discrete geometry developed by Hübler [Hü89]. The main idea is to define the Z^n as a neighborhood structure $[P,N]$ where translations are allowed (Hübler restricts his investigations to Z^2).

Def. 4.1.-1: A mapping T of P onto P is called *translation* if for any two points $p,q \in P$ with $(p,q) \in N$ also $(T(p),T(q)) \in N$, and for the graph distances the equation $d(p,T(p))=d(q,T(q))$ is fulfilled. We denote by T^{-1} the *inverse translation*, and the sign $\circ$ shall symbolize the connection of two translations. The expression T^k means the k-fold successive application of translation T.

Def. 4.1.-2: A neighborhood structure $[P,N]$ is called *translative* if the set $\boldsymbol{T}$ of all translation fulfills the following conditions:

- $\boldsymbol{T}$ is closed in respect of the operation $\circ$.
- The operation $\circ$ is commutative.
- For any points p,q with $(p,q) \in N$, there is a translation $T \in \boldsymbol{T}$ with $T(p)=q$.

- For all translations T different from the identical mapping $I=T^0$ and for any point $p \in P$, it does not exist integer numbers $c>0$ with $T^c(p)=p$.

Because of these definitions, the algebraical structure $[\boldsymbol{T},\circ]$ is an Abelian group. Further, finite neighborhood structures (toroidally closed Z^n) are not allowed to be translative.

As already mentioned at the end of section 2.3.2, we have to take into account finite (toroidally closed) or infinite ("planar") incidence structures Z^n with respect to the interesting actual problems. For geometrical investigations, the infinity of the structure is the suitable model. Because of the assumption that there is an inverse translation T^{-1} to each translation T, the number ν of neighbors of a point $p \in P$ has to be independent of p, and ν must be an even number.

Def. 4.1.-3: Translations T with translation difference $d(p,T(p))=1$ are called *elementary translations*. If a translation T cannot be represented as a power of any other $T' \in \boldsymbol{T}$, we will T denote as *simple translation*.

There are exactly ν different elementary translations in general, and $\nu/2$ elementary translations represent a generation system of $[\boldsymbol{T},\circ]$. Further, it follows that there exists a simple translation S and an integer $c \geq 1$ with $S^c=T$ for any translation T.

Def.4.1.-4: A subset $M \subseteq P$ of the neighborhood structure $[P,N]$ is *closed* in respect of translation T if for any $p \in M$ and any integer c also $T^c(p) \in M$. The smallest subset closed in respect of translation T which contains a given subset $M \subseteq P$ is denoted as *T-hull* of M.

Def.4.1.-5: A set $L \subseteq P$ of the neighborhood structure $[P,N]$ is a *discrete straight line* if L is the T-hull of any one-point set $\{p\}$ with $p \in P$, and T as a simple translation. T is the *generator* of L.

Each discrete straight line - shortly named *line* - can be generated by two generators T and T^{-1}. Such lines are generally no connected point sets (figure 4.1.-5). The generators for lines in Z^n are n-tupels of integers without common divisor. Therefore, for any two points $p=(x_1,\ldots,x_n)$ and $q=(y_1,\ldots,y_n)$, both elements of P, there is a unambiguously determined line with the generator

$$\frac{1}{c}(x_1-y_n, \ldots, x_n-y_n)$$

where c is the greatest common divisor of the coordinate differences x_i-y_i. Similar to Euclidean geometry, we can describe a line either in the form (p,T) by one point p and a generator T or in the form (p,q) by two points.

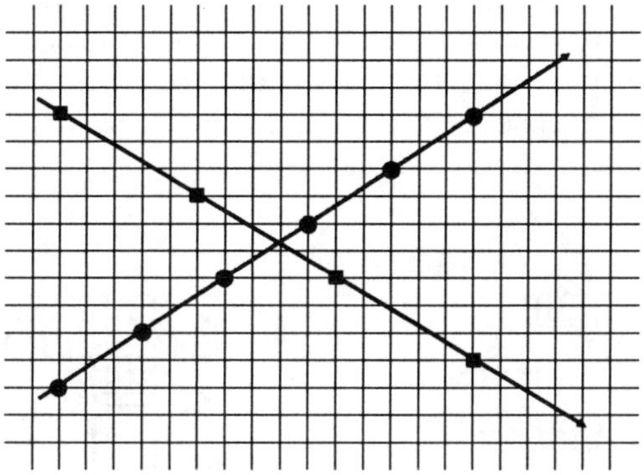

Fig. 4.1.-5: Two discrete straight lines with generators (3,2) and (5,-3)

Def. 4.1.-6: Two lines L and L' are *parallel* if they have the same generators. Two lines L and L' in Z^2 are *orthogonal* to one another if the scalar product of the generators T_L and $T_{L'}$ vanishes.

This short comments on the axiomatically based design of a discrete geometry shall be enough to illustrate Hübler's theory. Using the given definitions and the prototype of Hilbert's axiomatic method, Hübler also investigates further problems such as convexity, isomorphy, and finite parts of straight lines.

It shall be pointed out that we have to use cyclic Abelian groups of translations when we work with toroidal structures as homogeneous supports for discrete images (see chapter 5.1).

A related approach was given by Mylopoulos and Pavlidis already in 1971, but they restrict themself to an intensive investigation of dimensionality:

"Therefore it is necessary to introduce new definitions even for the most basic concepts of geometry. Actually we must go even further and introduce a formal definition of what we mean by a discrete or quantized space. Many authors assume that one deals with a 'checkerboard' situation...but this is certainly a rather arbitrary assumption and one can certainly consider other quantization schemes, e.g. a hexagonal grid. We follow a different approach and introduce an abstract definition.

Definition 1. A discrete space is a finitely presented Abelian group $G=(X/D)$, where X has $2n$ generators $s_1, s_2, \ldots, s_n$, s_1^{-1}, $s_2^{-1}, \ldots, s_n^{-1}$, and D contains all relations other than the commutativity ($s_i s_j s_i^{-1} s_j^{-1}=1$) and the inverse iterations ($s_i s_i^{-1}=1$)...Up to this point, our discussion is valid for any Abelian group. Now we restrict our attention to free Abelian groups, namely those groups that have no relations other than the commutativity and inversion relations... We will show that for such groups the dimension equals the number of generators...Thus there seems to be a correspondence between Euclidean spaces and free Abelian groups." [Mylopoulos and Pavlidis, My71]

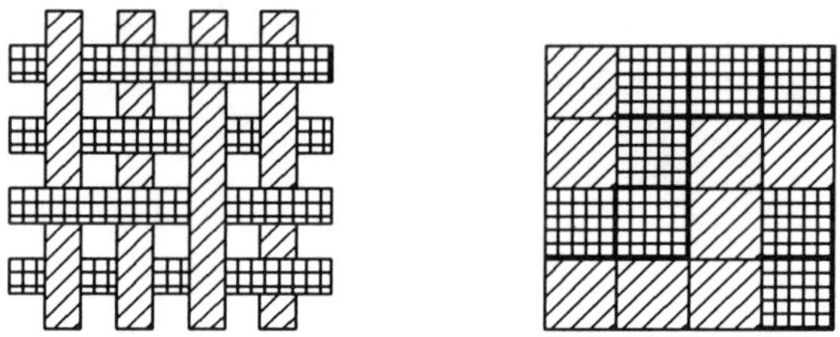

Fig. 4.1.-6: Textured web (left) and corresponding binary image (right)

An interesting problem for translative neighborhood structures was discussed by Grünbaum and Shephard [Gr85]. In figure 4.1.-6, a textured web is shown where the warps (lines in the corresponding digital image) and the wefts (columns in the image) are jointed together. Such a fabric is denoted as (k,l)-*isonemal* if there is a two-dimensional $k \cdot l$-periodicity, i.e. if there are periodical matching translations.

There is the question how many web structures are possible for given k and l. The answer is certainly not trivial. Problems arise even for small values of k and l as shown in figure 4.1.-7 for $k=l=3$.

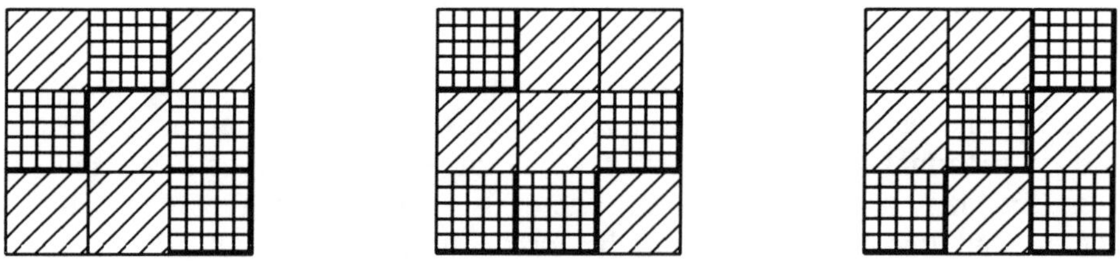

Fig. 4.1.-7: Three equivalent isonemal web structures

4.1.4 Digitalization Effects

The number of lattice points hitting an any shaped compact point set S in E^2 is an estimation of the area $A(S)$ of the set. More precisely, integral geometry shows that $A(S)$ is equal to the mean value of $N(S)$ with respect to all locations and orientations of a movable set S where N means the number of covered lattice points [Sa76]. The deviations of $N(S)$ from the mean value shall be called by the notion *digitalization effects*.

A very interesting approach to this problem was given by Klette [Kl83] with the following definition (see figure 4.1.-8).

Def. 4.1.-7: Let be $S \subset E^n$ a point set in the n-dimensional Euclidean space, $B = S \cap Z^n$ the set of all lattice points covered by S (i.e. a discrete object), and $D_k(B)$ with $0 \le k \le n$ the set of all k-dimensional object elements. Then $D_k(B)$ is called the k-dimensional *cell-digitalization* of S.

For convex sets S, we can define upper and lower bounds for the number $\varepsilon(S)$ of lattice points covered by S. The results reported in section 4.1.1 yield

$$V(C) - \frac{1}{2} S(C) \le \varepsilon(C) \le \sum_{i=0}^{2} \sum_{j=1}^{\binom{2}{i}} V_2\left(C_j^i\right)$$

for convex sets $S = C$. From this inequality, we obtain

$$A(C) - \frac{1}{2} P(C) \le \varepsilon(C) \le 1 + W_x(C) + W_y(C) + A(C) .$$

Here $A(C)$ and $P(C)$ mean area and perimeter of the convex point set C, respectively, and $W_x(C)$ and $W_y(C)$ are the widths of C in x-direction and y-direction, respectively (we have used $V_0(C_j^0) = 1$).

In image processing, the investigation of digitalization effects for any-shaped (two-dimensional) objects has been of main interest for many years [Яш65, Ba67]. At most, simulation experiments were performed to get hints and insights into statistical properties of occurring effects. An exact solution is only possible for one-dimensional problems [Vo75].

Let S be a segment of length l from point x to point $x+l$. $N_l(x)$ shall be the number of points with integer x-coordinates covered by S. Then $ent(x+l)$ is the greatest and $1+ent(x)$ is the smallest integer coordinate assuming that x is not integer. We obtain

$$N_l(x) = ent(x+l) - ent(x) .$$

The function $N_l(x)$ is periodical with period 1. Without loss of generality, we can assume that $0 < x < 1$ with $ent(x) = 0$. Using $d = l - ent(l)$ and

$$ent(x+l) = ent(x+ent(l)+d) = ent(l) + ent(x+d) \quad ,$$

it follows

$$N_l(x) = ent(l) + \begin{cases} 0 & \text{for} \quad 0 < x+d < 1 \\ 1 & \text{for} \quad 1 < x+d < 2 \end{cases} \quad .$$

The probability to get *ent(l)* points with respect to any location x is $p_0=1\text{-}d$, and the probability to get $1+ent(l)$ points is $p_1=d$. Therefore, we obtain for the expectation value of the number N_l of points covered by S and for the standard deviation

$$E(N_l) = l \quad , \qquad \delta N_l = \sqrt{d - d^2} \quad .$$

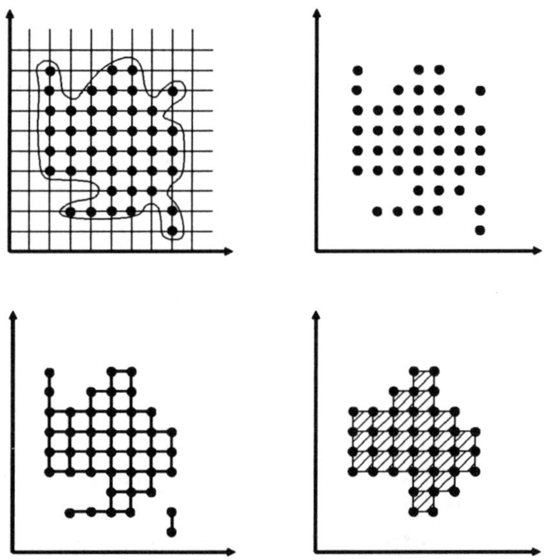

Fig. 4.1.-8: Digitalization results: $k=0$ (upper right), $k=1$ (lower left), and $k=2$ (lower right)

Now we assume that a two-dimensional convex set S shall be translated only in x-direction [Vo87b]. Then for each line parallel to the x-axis having an integer y-coordinate and hitting the set S, we obtain a chord with any length l. The digitalization effects in respect of this chord are discussed above.

In the orthogonal lattice, there are k of such chords. Because each of these chords gives a contribution $\Delta \in \{0,1\}$ to the total digitalization effect, we obtain

$$N_S(x) = N_{\min} + \Delta(x) \quad \text{with} \quad 0 < \Delta(x) < k \ .$$

Further, we introduce two assumptions:

- For large objects S, the mean values of probabilities p_0 and p_1, describing the contribution 0 and 1 to $\Delta(x)$, are 0.5.
- The single chord-dependent digitalization effects are largely independent of each other.

Then, we obtain a binomial distribution for the values $m=\Delta(x)$ with $0 \le m \le k$, namely

$$p_k(m) = 2^{-k} \binom{k}{m} \ .$$

The mean value of lattice points covered by S with respect to all x-translations is equal to the area $A(S)$ independent of the orientation of S.

The dispersion (i.e. the square of the standard deviation) of this $p_k(m)$-distribution is $D=k/4$. Because the expectation value of covered lattice points is independent from the orientation of S, the mean dispersion is determined by the expectation value of the number k of chords, i.e.

$$E(D) = E(k)/4 \ .$$

The expectation value $E(k)$ of chords of a convex two-dimensional set S with chord distance 1 is given by [Sa76]

$$E(k) = \frac{1}{\pi} P(S)$$

with $P(S)$ as perimeter of S. Therefore, we obtain

Formula 4.1.-1:

$$E(N_S) = A(S) \quad , \quad \delta N_S = \sqrt{E(D)} = \sqrt{\frac{P(S)}{4\pi}} \ .$$

The first equation is right for small and large, for convex and nonconvex sets S. The second equation is nearly exact for large convex sets with so many chords that the two assumptions (mean probabilities $p_0=p_1=0.5$ and chord-indepence) are valid.

For the simplest plane figure, the circle C with radius R, it is possible to determine δN_C directly. It was shown [Ke63] that

$$E(N_C) = \pi R^2 ,$$

$$E(N_C^2) = \pi^2 R^4 + R^2 \sum_{l=1}^{\infty} \frac{r(l)}{l} J_1^2(2\pi R\sqrt{l}) .$$

Here $J_1(x)$ is a first-kind Bessel function, and $r(l)$ means the number of all representations of l as sum of two squares of integers. Now we can determine δN_C by

$$\delta^2(N_C) = E(N_C^2) - \left(E(N_C)\right)^2 = R^2 \sum_{l=1}^{\infty} \frac{r(l)}{l} J_1^2(2\pi R\sqrt{l}) .$$

We have to estimate the infinite sum for further estimation of δN_C. An asymptotic expression of $J_1(x)$ is given by [Sm61]

$$J_1(x) \approx \sqrt{\frac{2}{\pi x}} \cdot \cos\left(x - \frac{3\pi}{4}\right) .$$

We obtain

$$\delta^2 N_C \approx \frac{2R^2}{2\pi^2 R} \sum_{l=1}^{\infty} \frac{r(l) \cdot \cos^2(l - 3\pi/4)}{l\sqrt{l}} .$$

Because the mean size of cos^2 is 1/2 , and the mean size of $r(l)$ can be estimated [Kr81] by π, it follows that

$$\delta^2 N_C \approx \frac{R}{2\pi} \sum_{l=1}^{\infty} \frac{1}{l^{3/2}} .$$

The infinite sum in this last formula is known as Riemann's zeta-function

$$\zeta(s) = \sum_{l=1}^{\infty} \frac{1}{l^s}$$

for $s=3/2$ with $\zeta(1.5)\approx 2.612$. Using $P(C)=2\pi R$ as perimeter of the circle, finally we obtain

$$\delta N_C \approx \sqrt{\frac{P(C)}{5.23\,\pi}} .$$

Perhaps it is possible to estimate δN_S also for other figures S than circles. But the statistically determined formula 4.1.-1 is more general.

As experiments show, the formula 4.1.-1 for δN_S is useful for any plane figures, also for nonconvex sets S as very often regarded in image processing. In the case of any sets S, we have to replace the mean chord number $E(k)$ by the half of the mean number $E(b)$ of intersections between lines with integer y-coordinate and the boundary of the set S [Sa76]:

$$E(b) = \frac{2}{\pi} P(S) \ .$$

The expectation value of chord number in convex and nonconvex sets is $E(k)=E(b)/2$. Therefore the same formula 4.1.-1 for δN_S follows. In figure 4.1.-9, some simulation results are represented to show the applicability of the here derived general formula 4.1.-1 of digitalization effects for plane figures.

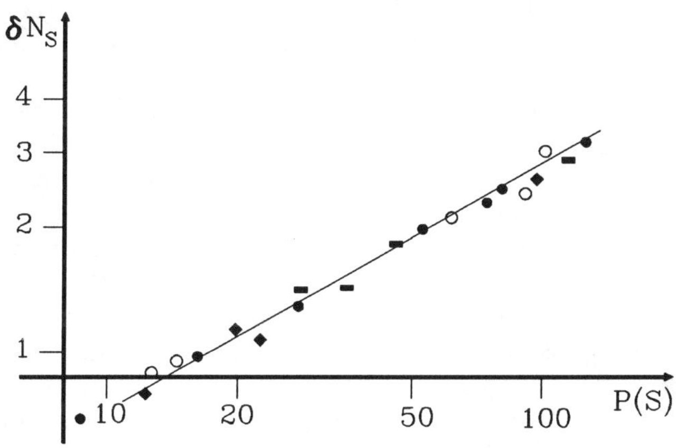

Fig. 4.1.-9: Results of simulation experiments for digitalization effects using circles (●), circular rings (○), squares (♦), rectangles (▬)

4.2 Straight Lines

4.2.1 Rational Geometry

The simplest geometrical objects are points and straight lines. Beginning with Euclid up to Hilbert's "Grundlagen der Geometrie", the properties and laws of these objects are the base for all further investigations and developments. As shown in section 4.1.3, the notion of straight line stands in the beginning also in discrete geometry.

Discrete geometry is - when we work with distances as metric notions - first of all a *rational geometry* in the sense that all coordinate values are integers and all linear relationships between integers lead to rational expressions. In this connection, also ideas of *finite geometries* shall be mentioned [Ka76, Kl85a].

A comprehensive view of problems connected with digital straight lines was given by Bruckstein who pointed out that modern demands on mathematical tools can be fulfilled some times by a search in very old literature:

> "It is very nice to see the many interesting connections between digitized straight lines and both classical and relatively new results in mathematics. As we have seen, questions implicitly concerning patterns appearing as chain-codes of digitized lines date back to the Greek mathematicians interested in measuring length or designing gears, and were also of interest to astronomers analyzing the interactions between sequences of occurrences of periodic events. As a result, various aspects of such problems never ceased to attract the interest of reasearchers in several fields of research." [Bruckstein, Br91]

It is not astonishing that discrete geometrical investigations lead to a number of theoretical problems in almost all cases. A small compilation of such tasks shall be given here:

- Given a small set of $M \cdot N$ points in the orthogonal square grid where the points (x,y) have integer coordinate values $x \in \{0,\ldots,M-1\}$ and $y \in \{0,\ldots,N-1\}$. How many different rational slopes $\Delta y/\Delta x$ exist?

- Which properties have grid point triangles without grid points inside of the triangle or at the boundary of the triangle?

- What is the upper limit of the number of corner points of a convex grid point polygon lying in the above defined $M \cdot N$ box?

- By which properties are grid point triangles characterized which have integer valued side lengths?

- Are there grid point tetrahedrons in Z^3 with six integer valued edge lengths?

To solve these problems, a large number of theoretical methods are to be exploited. Therefore, there is no simple unbranched way on which step by step results of the above stated tasks can be obtained. I think that only a lot of number theoretical knowledge and experience can be helpful for investigation and solution of problems in discrete geometry.

Certainly, not all number theoretical investigations and research efforts are necessary for image processing. But we are not sure that this is the case. Therefore, a narrow connection is desirable between discrete mathematics and image processing.

One of the useful mathematical disciplines is the theory of lattices. In [Co88] for instance, the properties of a lot of lattices are investigated including the cubic lattices Z^n which is also the theme of this book.

It can be shown that there are no equilateral triangles in Z^2: The area A of a triangle of lattice points is always a rational number

$$A = \frac{1}{2} \begin{vmatrix} x_1 & y_1 & 1 \\ x_2 & y_2 & 1 \\ x_3 & y_3 & 1 \end{vmatrix} ,$$

whereas the area A_e of a equilateral triangle is $s^2\sqrt{3}/4$. Therefore A_e is an irrational number because the square s^2 of the distance s between two lattice points is rational.

But there are equilateral triangles in Z^3, for instance the triangle given by the three points (0,0,0), (1,0,1), and (0,1,1) with point distance $s=\sqrt{2}$. There are also regular tetrahedrons in Z^3 as the example (0,0,0), (3,0,3), (0,3,3), (-1,-1,4) shows where $s^2=18$.

General simplexes in E^n as triangles in E^2 and tetrahedrons in E^3 were investigated by Rosenfeld and Jaglom [Ro71]. The volume V_n of a regular n-dimensional simplex with side length s is given by the recursion formula

$$h_1 = s \quad , \quad V_1 = s \, ,$$

$$h_n = \sqrt{h_{n-1}^2 - \frac{h_{n-1}^2}{n^2}} = \frac{h_{n-1}}{n} \sqrt{n^2-1} \, ,$$

$$V_n = \frac{V_{n-1} \cdot h_n}{n} \, .$$

Here, h_n is the height of the regular n-dimensional simplex, i.e. the distance of the center of gravity of n corner points from the $(n+1)$-th point. All heights of an equilateral simplex meet themselves in the center of gravity, and the center of gravity of an $(n-1)$-dimensional equilateral simplex divides each height with proportion $(n-1):1$. We can prove by induction that

$$h_n = \frac{s}{n}\sqrt{\frac{n(n+1)}{2}}, \quad V_n = \frac{s^n \sqrt{\prod_{k=1}^{n} \frac{k(k+1)}{2}}}{\prod_{k=1}^{n} k^2} .$$

Finally we obtain

$$V_n = \frac{s^n}{n!}\sqrt{\frac{n+1}{2^n}} = \frac{1}{n!}\sqrt{(n+1)\left(\frac{s^2}{2}\right)^n} .$$

It must be guarenteed that the expression within the square root must be a square of a rational number since the volume of each grid point simplex is a rational number. In general there is no solution for even values of $n=2m$ because

$$V_{2m} = \left(\frac{s^2}{2}\right)\frac{\sqrt{2m+1}}{(2m)!}$$

cannot be a rational number if not $2m+1$ is a square. For instance, we obtain with $m=4$ the nine points

$$\begin{array}{l}
(0, 0, 0, 0, 0, 0, 0, 0) \\
(s, s, 0, 0, 0, 0, 0, 0) \\
(s, 0, s, 0, 0, 0, 0, 0) \\
(s, 0, 0, s, 0, 0, 0, 0) \\
(s, 0, 0, 0, s, 0, 0, 0) \\
(s, 0, 0, 0, 0, s, 0, 0) \\
(s, 0, 0, 0, 0, 0, s, 0) \\
(s, 0, 0, 0, 0, 0, 0, s) \\
(\frac{s}{2}, \frac{s}{2}, \frac{s}{2}, \frac{s}{2}, \frac{s}{2}, \frac{s}{2}, \frac{s}{2}, \frac{s}{2})
\end{array}$$

which form an equilateral 8-dimensional simplex with side length l determined by $l^2=2s^2$.

4.2.2 Digital Straight Lines in Z^2

The problem of digital straight lines is one of the oldest problems in image processing [Fr61, Fr71]. Already in [Fr61], Freeman has given an interesting approach to two-dimensional curves encoded by sequences of direction codes (see figure 4.2.-1): In the case that there are some regularities in such a code sequence, the typical *sequence pattern* can be exploited to simplify the curve representation.

This simplification consists in use of *generating functions*. Be $S=(c_0c_1...c_K)$ the typical part of the code sequence repeated periodically, the corresponding generating function is defined by

$$G_S(x) = \sum_{k=0}^{K} c_k x^k .$$

For $S=(0,1)$, we obtain $G_S(x)=x$, and for the infinite sequence $S'=(010101...)$, it follows with $|x|<1$ the function

$$G_{S'}(x) = x + x^3 + x^5 + ... = \frac{x}{1 - x^2} .$$

Let L be the length of the pattern P of a curve C where P is J-fold repeated. Then

$$G_C(x) = G_P(x) \sum_{j=0}^{J} x^{jl} = G_P(x) \cdot \frac{1 - x^{JL}}{1 - x^L} .$$

As Freeman has already mentioned, the r-fold magnification ("expansion") of a curve C to curve C_r can be described by a transformation of the generating function $G_C(x)$ to

$$G_{C_r}(x) = \sum_i c_i \sum_{j=0}^{r-1} x^{ir+j} = \sum_i c_i x^{ir} \sum_{j=0}^{r-1} x^j$$

$$= \sum_i c_i x^{ir} \frac{1 - x^r}{1 - x} = \frac{1 - x^r}{1 - x} G_C(x^r) .$$

In 1974, Rosenfeld gave a first exact definition of digital straight line segments [Ro74]. The basic notion was the so-called chord property which generalizes the definition of convexity in E^2 (see also [Hü89]). Rosenfeld has proved that a simple two-dimensional ("irreducible") curve C with 8-neighborhood of curve points is then and only then a digital straight line segment if C has the chord property.

Def. 4.2.-1: A subset $C \subset Z^2$ has the *chord property* if for any two different points $p,q \in C$ the following condition is fulfilled: For each point r of the continuous line pq (embedded in E^2), there is a point $s \in C$ with

$$\max(|x(s) - x(r)|, |y(s) - y(r)|) < 1 .$$

In terms of the eight main directions in the orthogonal lattice Z^2 (see figure 4.2.-1), a digital straight line with rational slope $\Delta y/\Delta x$ smaller than 1 can be coded by a sequence of 0's and 1's. For the successive order of the code numbers, Freeman has conjectured that [Fr71]

i) only two consecutive codes (here 0 and 1) are used in the code sequence,
ii) at least one of these two direction codes is always isolated,
iii) successive occurrences of the isolated direction code are as uniformly spaced as possible.

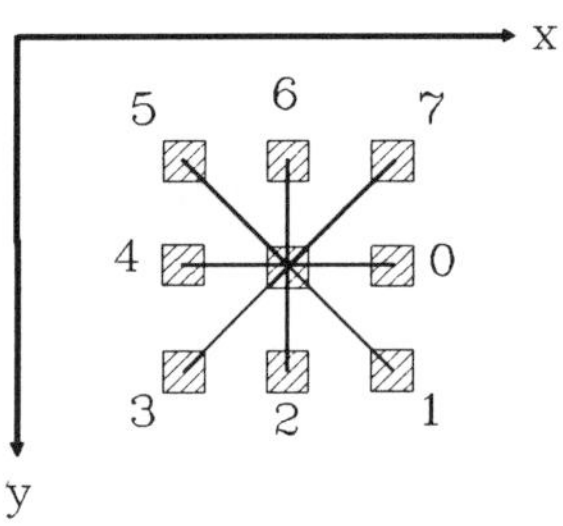

Fig. 4.2.-1: Direction codes

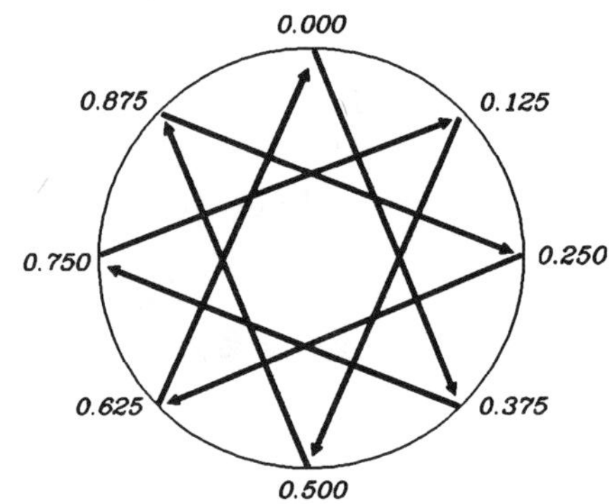

Fig. 4.2.-2: Spirograph of a straight line $y=ax+b$ for $a=0.125$

Later, Rosenfeld has shown by use of the *chord property* that the third conjecture can be expressed exactly as follows:

iv) The run length of the non-isolated direction code fulfills the conditions i), ii), and iv).

This result shows that a recursive method maybe could give a simple theoretical approach to the problem of description of digital straight lines.

A very clear representation of the properties of straight lines in Z^2 is given by the so-called *spirograph* theory [Do84, Do89]: A straight line shall be determined by equation $y=ax+b$. The sequence of "rounding differences"

$$\alpha = 2\pi (y - ent(y))$$

for $\Delta x=1$ and $b=0$ is interpreted as a sequence of angles or arcs of a circle (figure 4.2.-2). For rational values $a=k/l$ with $gcd(k,l)=1$, this sequence is periodical, i.e. the spirograph is a general polygon with l corners.

Dorst has developed a full theory of spirographs and has it applied to the problem of digital straight lines [Do86]. But there are also other approaches to the interesting properties of digitized straight lines. Such a theoretical framework based on continued fractions will be given in the next section [Vo91b].

In [Ro76a], it was discussed that digital line segments of length L through the origin, i.e. line segments corresponding to equation $y=\alpha x$ with $0\leq\alpha\leq 1$, are in one-to-one correspondence with the *Farey sequences* of length L (see section 4.3.2). In respect of this, it follows from a well-known result in number theory that the number N_L of line segments of lengths l with $1\leq l\leq L$ through the origin is

$$N_L = \sum_{l=1}^{L} \varphi(l) \approx \frac{3L^2}{\pi^2} + O(L\cdot\log L) .$$

Here, $\varphi(l)$ is *Euler's function*, i.e. the number of all numbers k with $1\leq k\leq l$ which have no common divisors with l. For instance, it is

$$N_6 = \varphi(1)+\varphi(2)+\varphi(3)+\varphi(4)+\varphi(5)+\varphi(6) = 1+1+2+2+4+2 = 12 .$$

In [Be88] it was shown that the number of different slopes of line segments with integer valued end point coordinates within an $L\cdot L$-picture can be approximated by $4\cdot N_L$.

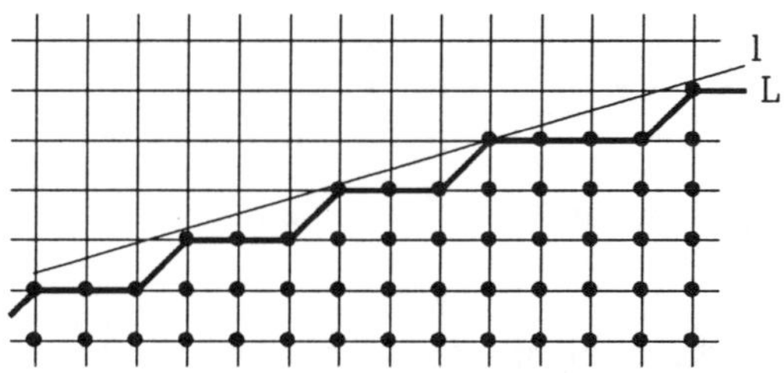

Fig. 4.2.-3: A line l and its digital picture L

4.2.3 Continued Fractions

To shorten the argumentation, we restrict generally our investigations to lines with a rational slope smaller than 1 which go through at least two lattice points (this condition can be always fulfilled by a simple translation of the line). We will denote such lines as *rational lines*. The digital picture of a rational line l is the set L of all boundary points which are not above the line l (figure 4.2.-3).

The slope of a rational line l is determined by a tupel $(\Delta x, \Delta y)$ of integers where the greatest common divisor $gcd(\Delta x, \Delta y)$ is equal to 1. The differences Δx and Δy determine the generator of a discrete line (see section 4.1.3), and the triangle given by the points $(x,y),(x+\Delta x,y)$, $(x,y+\Delta y)$ is the *characteristical triangle* of the corresponding line (figure 4.2.-4). The result of the *concatenation* $T'\oplus T''$ of two characteristical triangles T' and T'' are once more a characteristical triangle T with

$$\left(\Delta x_T, \Delta y_T\right) = \frac{1}{c}\left(\Delta x_{T'} + \Delta x_{T''}, \Delta y_{T'} + \Delta y_{T''}\right)$$

whereby c is such an integer that $gcd(\Delta x_T, \Delta y_T)=1$.

Def. 4.2.-2: A rational straight line segment pq shall be denoted as *dissectible* in two other straight line segments pr and rq if there are no lattice points inside of the triangle pqr.

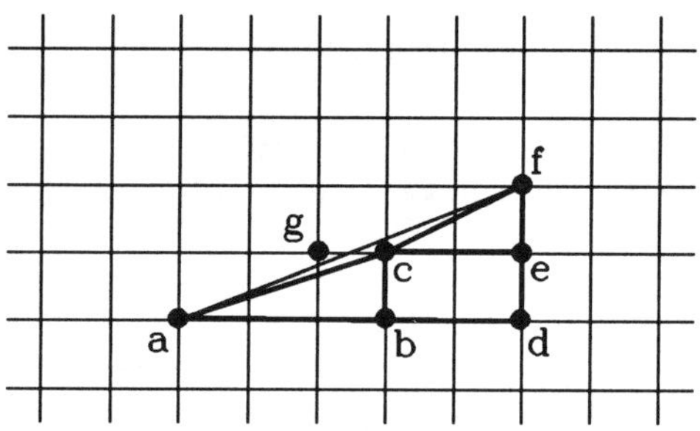

Fig. 4.2.-4: Two digital straight line segments *ac* with slope 1/3 and *cf* with slope 1/2

In figure 4.2.-4, the straight line segment *af* is dissectible because the concatenation of the both line segments *ac* and *cf* yields the same tupel (5,2) as segment *af* whereby the triangle *acf* contains no lattice points.

We can use Pick's formula $A=i+b/2-1$ for grid point triangles (see section 2.3.3). Because $i=0$ and $b=3$, the area A of all such triangles without inner points are always 1/2.

Now, we assume that the slopes of all lines and segments to be investigated are smaller than 1 and greater than 0, i.e. $\Delta x > \Delta y > 0$. In this case, the rational number $\Delta y/\Delta x$ can represented as a finite *continued fraction* [Sc91]

$$\frac{\Delta x}{\Delta y} = [a_1, a_2, \ldots, a_n] = \cfrac{1}{a_1 + \cfrac{1}{a_2 + \cfrac{1}{\ddots \cfrac{1}{a_{n-1} + \cfrac{1}{a_n}}}}}$$

with rational coefficients $a_i > 0$.

The continued fractions of this type have some interesting properties. At first, it is easy to prove that $[a_1+1] < [a_1]$ and $[a_1, a_2+1] > [a_1, a_2]$. In the general case, we obtain (provable by induction)

$$[a_1, \ldots, a_n+1] < [a_1, \ldots, a_n] \qquad \text{for odd } n ,$$

$$[a_1, \ldots, a_n+1] > [a_1, \ldots, a_n] \qquad \text{for even } n .$$

The numerical value of a finite continued fraction is always a common fraction

$$[a_1, \ldots, a_n] = \frac{\alpha_n a_n + \beta_n}{\gamma_n a_n + \delta_n}$$

where the expressions $\alpha_n, \beta_n, \gamma_n, \delta_n$ are functions of the coefficients $a_1, \ldots, a_{n-1}$. For $n=1$, we obtain

$$\alpha_1 \delta_1 - \beta_1 \gamma_1 = -1 .$$

Further, it is

$$[a_1, \ldots, a_{n-1}, a_n, a_{n+1}] = \frac{\alpha_{n+1} a_{n+1} + \beta_{n+1}}{\gamma_{n+1} a_{n+1} + \delta_{n+1}}$$

$$= \left[a_1, \ldots, a_{n-1}, a_n + \frac{1}{a_{n+1}}\right] = \frac{\alpha_n \left(a_n + \frac{1}{a_{n+1}}\right) + \beta_n}{\gamma_n \left(a_n + \frac{1}{a_{n+1}}\right) + \delta_n}$$

$$= \frac{\alpha_n (a_n a_{n+1} + 1) + a_{n+1} \beta_n}{\gamma_n (a_n a_{n+1} + 1) + a_{n+1} \delta_n} .$$

Therefore, it follows

$$\frac{\alpha_{n+1} a_{n+1} + \beta_{n+1}}{\gamma_{n+1} a_{n+1} + \delta_{n+1}} = \frac{(\alpha_n a_n + \beta_n) a_{n+1} + \alpha_n}{(\gamma_n a_n + \delta_n) a_{n+1} + \gamma_n} .$$

Finally, we obtain

$$\alpha_{n+1}\delta_{n+1} - \beta_{n+1}\gamma_{n+1} = -(\alpha_n\delta_n - \beta_n\gamma_n) ,$$

and the general relationship

$$\alpha_n\delta_n - \beta_n\gamma_n = (-1)^n$$

follows. The slope of a characteristical triangle determined by the corresponding continued fraction $[a_1,\ldots,a_n]$ can be transformed:

$$\frac{(\alpha_{n-1}a_{n-1} + \beta_{n-1})\,a_n + \alpha_{n-1}}{(\gamma_{n-1}a_{n-1} + \delta_{n-1})\,a_n + \gamma_{n-1}} =$$

$$= \frac{(\alpha_{n-1}a_{n-1} + \beta_{n-1})(a_n - 1) + \alpha_{n-1}(a_{n-1} + 1) + \beta_{n-1}}{(\gamma_{n-1}a_{n-1} + \delta_{n-1})(a_n - 1) + \gamma_{n-1}(a_{n-1} + 1) + \delta_{n-1}} .$$

The last expression is the slope which results by concatenation $\oplus$ of one triangle with slope $s_1=[a_1,\ldots,a_{n-2},a_{n-1}+1]$ and a_n-1 triangles of slope $s_0=[a_1,\ldots,a_{n-2},a_{n-1}]$. Whether first the one triangle with slope s_1 or first the a_n-1 triangles with slope s_0 have to be arranged is dependent on the requirement that the "left" slope must be smaller than the "right" slope (in figure 4.2.-4, the value 1/3 is smaller than 1/2, and therefore the point g is above the line af). We can scribe symbolically the last result as *splitting formula* of slopes:

Formula 4.2.-1:

$$[a_1,\ldots,a_n] = [a_1,\ldots,a_{n-1}+1] \oplus (a_n - 1)\cdot[a_1,\ldots,a_{n-1}] .$$

As an example, we will split the slope 68/157=[2,3,4,5] into elementary slopes $[a]=1/a$. It is

$$\begin{aligned}
[2,3,4,5] &= [2,3,5] \oplus 4\cdot[2,3,4] \\
&= (4\cdot[2,3]\oplus[2,4]) \oplus 4\cdot(3\cdot[2,3] \oplus [2,4]) \\
&= (4\cdot([3]\oplus 2\cdot[2]) \oplus ([3] \oplus 3\cdot[2])) \oplus \\
&+ 4\cdot(3\cdot([3] \oplus 2\cdot[2]) \oplus ([3] \oplus 3\cdot[2])) .
\end{aligned}$$

The splitting arrangement is performed in such a manner that alternatively the single slope s_0 or the multiple slope s_1 are the first. The elementary slopes $[a]$ with integer values of a can be easily coded by $a-1$ 0's and one 1. Therefore, the digital straight line segment with slope 68/157 can be coded by the following sequence:

```
0010101 0010101 0010101 0010101    001010101
        0010101 0010101 0010101    001010101
        0010101 0010101 0010101    001010101
        0010101 0010101 0010101    001010101
        0010101 0010101 0010101    001010101
```

Here 157 codes (157−68 zeros and 68 ones) are arranged in such a manner that "the successive occurrences of the isolated direction code are as uniformly spaced as possible".

Now, we can also prove Freeman's conjecture and Rosenfeld's requirement in an exact way. For even n, we obtain

$$
\begin{aligned}
[a_1,\dots,a_n] &= [a_1,\dots,a_{n-1}+1] \oplus (a_n-1)\cdot[a_1,\dots,a_{n-1}] \\
&= (a_{n-1}\cdot[a_1,\dots,a_{n-2}] \oplus [a_1,\dots,a_{n-2}+1]) \oplus \\
&\quad (a_n-1)\cdot((a_{n-1}-1)\cdot[a_1,\dots,a_{n-2}] \oplus [a_1,\dots,a_{n-2}+1]) \ .
\end{aligned}
$$

At this level, $[a_1,\ldots,a_{n-2}+1]$ is the singular symbol (i.e. at the first level the isolated direction code), and $[a_1,\ldots,a_{n-2}]$ is the repeated symbol. There are no other symbols, and the lengths a_{n-1} and $a_{n-1}-1$, respectively, differ by 1.

A periodical code sequence which fulfills the conditions i)...iv) can be used for definition of a digital straight line. But any admissible sequence of points is not sufficient to determine unambiguously a digital straight line. For instance, the point sequence ((0,0),(1,0),(2,1)) with code sequence (0,1) can be a part of a line with periodical code sequence (0,1) or it can be a part of a line with periodical code sequence (0,1,0,0,1) etc.

At the end of this section, we will once more quote Bruckstein who has used also continued fractions for description of digital straight lines [Ro76a, Ir89]:

> "Interestingly, the same problem was adressed simultaneously, in the mathematical literature, as a number theoretic question: when is a finite sequence of integers, $\{h_i\}$, representable as $h_i=[mi+n]$ for some numbers m,n. Such questions apparently date back to the astronomer among the Bernoulli brothers, being motivated by a computational question in astronomy...As it often happens in research, the number theorists were completely unaware of similar results published by their colleagues in computer vision, and vice versa." [Bruckstein, Br91]

4.2.4 Straight Lines in Z^n

The number lattice Z^n is a translative neighborhood structure. Therefore, each n-tupel $a=(a_1,\ldots,a_n)$ with integer a_i and $gcd(a_1,\ldots,a_n)=1$ is generator of a discrete straight line. We assume that the Z^n is embedded in the Euclidian space

E^n. Then both lines, the discrete line generated by a and determined by the lattice point x and the straight line

$$x_i = x_i^{(0)} + \tau \cdot a_i \qquad \text{for } \tau \in R$$

go through the same lattice points. But the discrete straight line L_Z is intuitively not the digital image of the continuous straight line L_E because too few lattice points are given by the discrete line in general.

There are different digitization schemes thinkable to yield a digital image L_d of L_E which has the following properties:

- L_Z is a subset of L_d,
- the projection of L_d at subspaces Z^m of Z^n with $m<n$ shall give a digital straight line also in that subspaces,
- the digitization rule has to be given for all dimensions.

Only very few investigations were performed in the past for digital images of m-dimensional linear objects (straight lines, planes, hyperplanes) in n-dimensional Euclidean spaces for $m>1$ and $n>2$. We will speak in this connection of *digital (m,n)-objects*: The digital straight lines investigated in section 4.2.3 are (1,2)-objects, and Kim has investigated lines and planes in Z^3, i.e. (1,3)-objects [Ki81] and (2,3)-objects [Ki82a], respectively.

> "The following two problems are among the many with which digital image processing is concerned. One is the problem of digitization, that is, how to represent continuous objects using finite sets of digital points. The other is the problem of retrieving from a set of digital points information about the object represented by it. Consider a line segment, for example. One problem is how to represent it by a set of digital points, and another is how we can tell whether or not a given set of digital points represents a line segment." [Kim, Ki81]

Rosenfeld and Melter have given a rating classification of the *digitalization problem* [Ro87], i.e. a classification of methods how to define digital objects $<S>$ from their continuous preimages S:

> "*Bad definition*: It might seem natural to simply define $<S>$ as the set of lattice points contained in S; but this means that even a 'large' S...could have an empty digitization...
>
> *Sloppy definition*: A better idea is to define $<S>$ as the set of lattice points that lie at distance $\leq 1/2$ from S...let S consists of a single point P; then $<S>$ may consists of 0,1 or 2 points, depending on the coordinate of P...

Less sloppy definition: We can do somewhat better if we use 'city block distance' instead of Euclidean distance...then $<S>$ can never be empty if S is nonempty...

Acceptable definition: .... all we need to do is to define $<P> \equiv <(i+\alpha,j+\beta)>$ to be the lattice point whose coordinates are obtained by *rounding* the numbers $i+\alpha$ and $j+\beta$ in some standard way...when α or $\beta=\frac{1}{2}$, we shall round downward." [Rosenfeld and Melter, Ro87]

The "bad definition" is used in the theory of numbers (see section 4.1.1) and for investigation of digitalization effects (section 4.1.4). Also in section 4.2.2, this definition was used when we have defined a digital straight line in Z^2 to be the set of all boundary points of a half-plane. But for (1,3)-objects, we have to work with the "acceptable definition" because the model of a straight line in Z^3 as intersection of two (2,3)-objects fails.

Def. 4.2.-3: Given a tupel $(a_1,...,a_n,m)$ of integer numbers a_i and m with $m \geq max(|a_i|) > 0$. The n-dimensional *digital straight line* going through the lattice point $p=(p_1,...,p_n)$ is defined by all points $q=(q_1,...,q_n)$ with coordinates

$$q_i = p_i + ent\left(\frac{t}{m} a_i\right) \qquad \text{for} \quad i \in \{1,...,n\}$$

for any integer t.

This definition gives the natural generalization of the notion of a digital straight line in Z^2 to Z^n. All dicussions were focussed in the two-dimensional case to the problem in which order 0's and 1's have to be arranged by stepwise enlargement of t. Now in Z^n, we have the same problem. A step $\Delta t=1$ yields for the q_i always

$$\Delta q_i = ent\left(\frac{t+1}{m} a_i\right) - ent\left(\frac{t}{m} a_i\right) \in \{-1,0,+1\} .$$

Up to reflections and symmetries, we can assume, that all a_i are non-negative integers. Then we have to investigate the arrangement of 0's and 1's in the Δq_i-sequences. Thereby it is sufficient to describe only one period of these sequences whereby a period has the length m.

As example, we investigate the three-dimensional digital straight line going through the origin and determined by the tupel $(a_1,a_2,a_3,m)=(7,11,3,13)$. The following table 4.2.-1 gives the coordinate values of the first period.

Table 4.2.-1: Coordinates and coordinate differences of a (1,3)-object

t	q_1	Δq_1	q_2	Δq_2	q_3	Δq_3
0	0		0		0	
1	0	0	0	0	0	0
2	1	1	1	1	0	0
3	1	0	2	1	0	0
4	2	1	3	1	0	0
5	2	0	4	1	1	1
6	3	1	5	1	1	0
7	3	0	5	0	1	0
8	4	1	6	1	1	0
9	4	0	7	1	2	1
10	5	1	8	1	2	0
11	5	0	9	1	2	0
12	6	1	10	1	2	0
13	7	1	11	1	3	1

The projections $(q_1,q_2)=(x,y)$, $(q_2,q_3)=(y,z)$, and $(q_1,q_3)=(x,z)$ of this three-dimensional digital straight line are represented in figure 4.2.-5. It is to be seen that the projections of a (1,3)-object are not necessarily (1,2)-objects in general. Therefore, it will be a task of the future to investigate digital linear (m,n)-objects in a full theory.

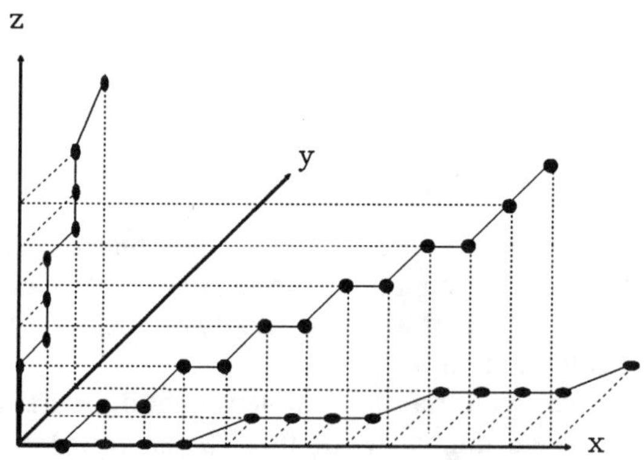

Fig. 4.2.-5: Projection of line points to the coordinate planes

4.3 Convexity

4.3.1 Convexity in Discrete Geometry

To define the notion of a *convex object* in Z^n, we have to use a so much as possible general model of convexity. In Euclidean geometry, the following approach is known to describe the convexity of a point set: A point set $M \subset E^n$ is called convex if for any two points $p,q \in M$ all points r of the straight line between p and q belong likewise to M.

But there are no further lattice points between two lattice points $p=\{0,0\}$ and $q=\{4,5\}$ although the set $Q=\{(p,q)/p,q \in \{0,1,...5\}\}$ is "obviously" a convex set. We have to find a definition of convexity which is related to Z^n and has analogies to Euclidean geometry.

Def. 4.3.-1: Three points $p_1=(x_1,y_1)$, $p_2=(x_2,y_2)$, $p_3=(x_3,y_3)$ of Z^2 are called *collinear* if the area of the triangle determined by the points is

$$A = \begin{vmatrix} x_1 & y_1 & 1 \\ x_2 & y_2 & 1 \\ x_3 & y_3 & 1 \end{vmatrix} = (x_2-x_1)(y_3-y_1) - (x_3-x_1)(y_2-y_1) = 0 .$$

If $A \neq 0$, we denote the points as *non-collinear*. This definition is legitimate both in discrete and in Euclidean geometry because the coordinate values can be real numbers, rational numbers, or integers. In this way, we have to define also convexity [Si75, Vo91].

Def. 4.3.-2: Let be $p_1,p_2,p_3 \in Z^2$ non-collinear points. The coordinates of any point $p=(x,y)$ can be represented by

$$x = x_1 + \lambda(x_2-x_1) + \mu(x_3-x_1) ,$$

$$y = y_1 + \lambda(y_2-y_1) + \mu(y_3-y_1) ,$$

$$\lambda = \frac{(x-x_1)(y_3-y_1) - (y-y_1)(x_3-x_1)}{(x_2-x_1)(y_3-y_1) - (y_2-y_1)(x_3-x_1)} ,$$

$$\mu = \frac{(x-x_1)(y_2-y_1) - (y-y_1)(x_2-x_1)}{(x_3-x_1)(y_2-y_1) - (y_3-y_1)(x_2-x_1)} .$$

The point p is *triangular* in respect of p_1,p_2,p_3 if $\lambda \geq 0, \mu \geq 0, \lambda+\mu \leq 1$.

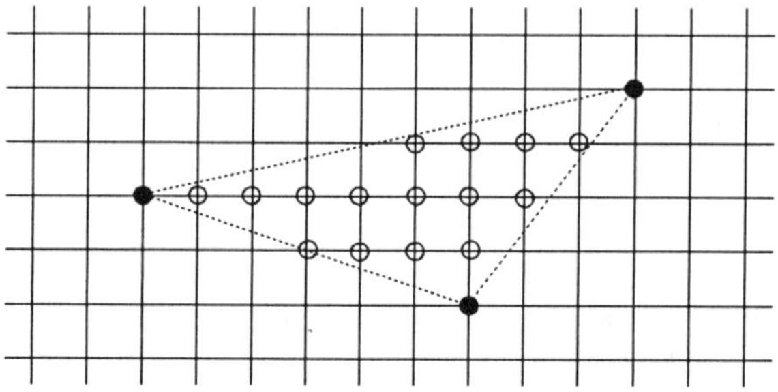

Fig. 4.3.-1: Triangular points (○) with respect to three lattice points (●)

The advantage of this somewhat cumbersome definition of triangularity is that it can be used also in both geometries. The difference to Euclidean geometry is that there all points with $\lambda \geq 0$, $\mu \geq 0$, $\lambda + \mu \leq 1$ are points of the (convex) triangle given by p_1, p_2, p_3. In discrete geometry, we have to investigate which lattice points are inside of the triangle or at the sides of the triangle (figure 4.3.-1). For points p_1, p_2, p_3 with finite coordinate values, there is always a finite number of triangular lattice points in respect of these three points. Using the notion of triangularity, we can formulate the notion of a convex lattice point set in Z^2.

Def. 4.3.-3: A set $B \subset Z^2$ is called *convex* if for any three points $p_1, p_2, p_3 \in B$, all points p which are triangular to p_1, p_2, p_3 belong likewise to B.

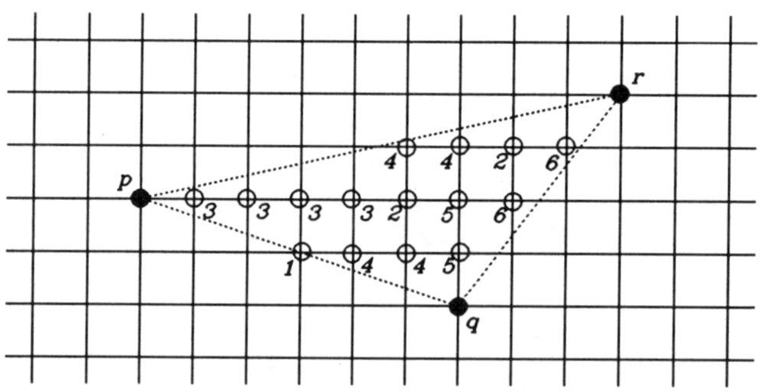

Fig. 4.3.-2: Stepwise construction of a convex set by means of closeness with respect to relation Z (see text)

There are other definitions of convexity which allow to develop corresponding geometrical theories [Sk70, Ga77, Ki82] where also the *chord property* plays an essential role (see definition 4.2.-1).

Hübler introduces a relation Z for three points p,q,r of a discrete straight line G in two-dimensional translative neighborhood structures (see section 4.1.3): It is $Z(p,r,q)$ fulfilled if $r=T^{j}(p)$, $q=T^{k}(p)$ and $0<j<k$ with T as generator of G [Hü89]. In this case, point r lies "between" points p and q. Then the closeness of a set B with respect to relation Z is introduced when from $p,q\epsilon B$ and $Z(p,r,q)$ also $r\epsilon B$ follows. Finally, a set B is called *convex* if B is closed with respect to the relation Z.

In figure 4.3.-2, a stepwise construction of a set B closed with respect to relation Z is represented. In this case, all points triangular to corner points p,q,r are yielded. But Hübler has pointed out in this connection the following:

> "Interessant ist die Feststellung, daß es außer 'linearen' konvexen Mengen wie Geraden, Strahlen und Strecken auch 'höherdimensionale' konvexe Punktmengen gibt, die nicht zusammenhängend sind...deshalb sollte man Möglichkeiten suchen, den Konvexitätsbegriff zu verschärfen." [Hübler, Hü89] [T7)]

The discreteness of Z^n opens the possibility for many interesting problems which are meaningless or cannot be defined in E^n. An essential reason for that is the fact that the elements of discrete spaces can be arranged by their neighborhood and can be counted.

In [Go85], a lot of similar problems are given which could be entitled by "complete order is impossible". For instance, Graham investigated problems of Ramsey theory [Gr85a]. An example for that is the theorem that each arrangement of numbers $k\epsilon\{1,2,\ldots,N^2+1\}$ contains always a monotone (increasing or decreasing) sequence of $N+1$ numbers.

Moser compiles some two-dimensional problems [Mo85]. His 13-th problem is characterized by the following: Be given N points in the plane where all tripels of points are not collinear. How large is $c(N)$ so that always a maximal convex polygon with c corners arises? It is $c(4)=3$, $c(5)=4$, $c(9)=5$, and there is the conjecture that $c(2^{K-2}+1)=K$.

4.3.2 Maximal Convex Objects

The following section is also one of the problems of this type. For a region R of m^2 points in Z^2, we investigate the problem how large is the maximal possible number $c(m)$ of corners of a convex object $B\subseteq R$, and how large is the minimal size $m(c)$ of a square-like region with c as given number of corners of a convex object [Vo82].

In figure 4.3.-3, maximal convex objects are represented for $m=8,9,10$. We will derive for two infinite subsets of integers c and m the exact values $c(m)$ and $m(c)$, and give two empiric approximation formulas for all values of c and m.

For small values of m and c, the function $c(m)$ and $m(c)$ may be determined using some kind of systematic search (see table 4.3.-1). Doing this, one gets very quickly an intuitive feeling about the complexity of such an approach. But in any case, the knowledge of formulas for $c(m)$ and $m(c)$ would offer the fastest way for computing those values.

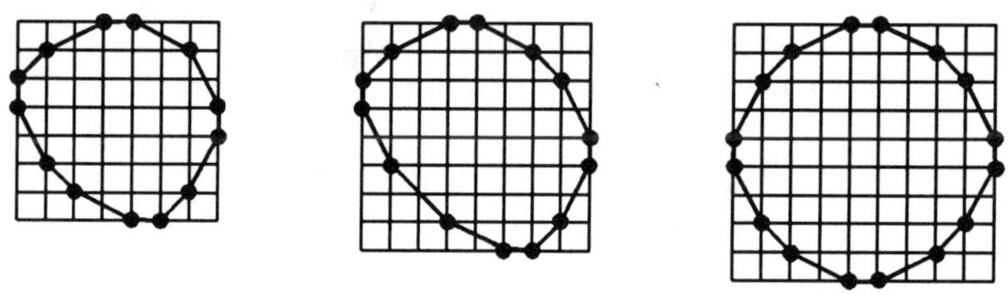

Fig.4.3.-3: Maximal convex objects with $c(8)=13$, $c(9)=14$, $c(10)=16$

Table 4.3.-1: Function values for $c(m)$ and $m(c)$

	1	2	3	4	5	6	7	8	9	10	11	12	13
$m(c)$	1	2	2	2	3	3	4	4	5	6	7	7	8
$c(m)$	1	4	6	8	9	10	12	13	14	16	16	17	18

We will use here an important notion of number theory, namely Euler's φ-function which gives the number $k=\varphi(n)$ of integers j which are not greater than n and relative prime to n, i.e. $gcd(j,n)=1$ with gcd as greatest common divisor of j and n [Kr81]. For instance, it is $\varphi(2)=1$, $\varphi(3)=2$, $\varphi(4)=2$, $\varphi(5)=4$, $\varphi(6)=2$, $\varphi(7)=6$. There is a very simple expression $\varphi(p)=p-1$ if p is a prime number. For powers of prime numbers is

$$\varphi(p^j) = p^j\left(1 - \frac{1}{p}\right) .$$

Now, we construct maximal convex objects with c corners for given square-like regions with m^2 lattice points. These objects may be characterized by closed sequences of finite lines between lattice points on the border of polygons where no three consecutive corner points are collinear (see figure 4.3.-3). Such straight line segments between corner points are the basic elements for building convex lattice point polygons. Without regard to the four possible directions of these lines, they can be described by tupels $(\Delta x, \Delta y)$ whereby $\Delta x, \Delta y > 0$ and $gcd(\Delta x, \Delta y) = 1$, or $(\Delta x, \Delta y) = (0,1)$, or $(\Delta x, \Delta y) = (1,0)$.

We investigate only "symmetrical" polygons as the third example in figure 4.3.-3. Then the basic sequence of lines for a given number t is $(t,1),(t-1,1), \ldots(\Delta x, \Delta y), \ldots(t-1,t),(1,1)$ where $gcd(\Delta x, \Delta y) = 1$ and $\Delta x > \Delta y$ (see figure 4.3.-4 for $t=4$). The single lines $(\Delta x_i, \Delta y_i)$ must be arranged in such an order that

$$\frac{\Delta y_{i+1}}{\Delta x_{i+1}} > \frac{\Delta y_i}{\Delta x_i} .$$

For $t=4$ we have for the first quadrant the sequence

$$\frac{0}{1} < \frac{1}{4} < \frac{1}{3} < \frac{1}{2} < \frac{2}{3} < \frac{3}{4} < \frac{1}{1} < \frac{4}{3} < \frac{3}{2} < \frac{2}{1} < \frac{3}{1} < \frac{4}{1}$$

so that there are 48 polygonal edges in total (i.e. four times the number of terms in this sequence).

Sequences of rational numbers of this type are called *Farey-sequences* [Sc91]. The N-th Farey-sequence $F^{(N)}$ consists of all rational numbers a/b with $gcd(a,b) = 1$ between 0 and 1 arranged in increasing order where the divisor is not greater than N:

$$F^{(5)} = \left(\frac{0}{1}, \frac{1}{5}, \frac{1}{4}, \frac{1}{3}, \frac{2}{5}, \frac{1}{2}, \frac{3}{5}, \frac{2}{3}, \frac{3}{4}, \frac{4}{5}, \frac{1}{1}\right) .$$

For two consecutive numbers a/b and a'/b' in such a sequence, we have always

$$a'b - ab' = 1 \qquad \text{for} \qquad \frac{a}{b} < \frac{a'}{b'} .$$

In general, we obtain for the number c_t of polygon edges, i.e. the number of polygon corners, the formula

$$c_t = 4\left[1 + \sum_{j=2}^{t} \varphi(j) + 1 + \sum_{j=2}^{t} \varphi(j)\right] = 8 \sum_{j=1}^{t} \varphi(j) .$$

The general expression for the size m_t of the constructed convex polygon is more complicated. We have to sum all Δx and all Δy in the sequence and to multiply this result by 2:

$$m_4 = 2\,[\,(1+4+3+2+3+4) + 1 + (3+2+1+1+1+0)\,] = 52 .$$

Now, we use the symbol $d|j$ to describe the fact that d is a divisor of j. We obtain

$$m_t = 2\left[\sum_{j=1}^{t} j\cdot\varphi(j) + 1 + \sum_{j=2}^{t}\sum_{\substack{k=1\\|k,j|=1}}^{j} k\right] .$$

The second sum means the sum over all k's for which the greatest common divisor with j is $gcd(k,j) = |k,j| = 1$. It follows

$$\sum_{\substack{k=1\\|k,j|=1}}^{j} k = \sum_{k=1}^{j} k - \sum_{\substack{k=1\\|k,j|>1}}^{j} k = \frac{j(j+1)}{2} - \sum_{\substack{d=1\\d|j}}^{j-1} \frac{j}{d} .$$

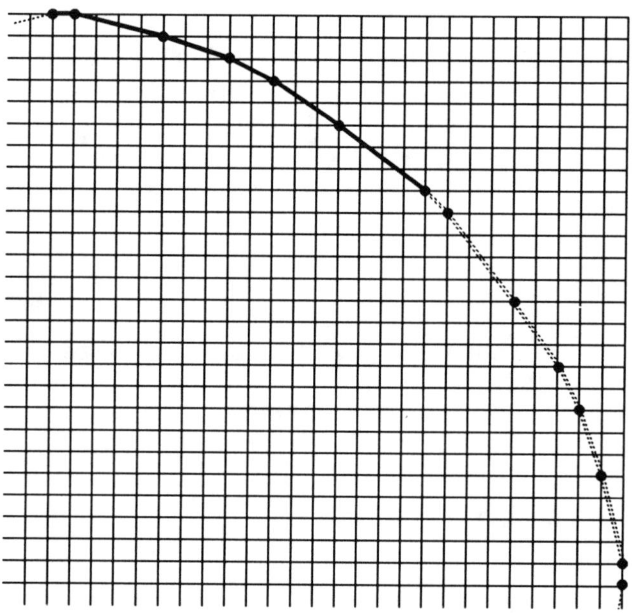

Fig. 4.3.-4: Part of a maximal convex polygon

Here the summation on the right side sum runs over all divisors d of j which are smaller than j. Because j/d is also a divisor of j if $d|j$, we obtain

$$\sum_{\substack{d=1\\d|j}}^{j-1} \frac{j}{d} = \sum_{\substack{r=2\\r|j}}^{j} r = \sum_{r|j} r - 1 .$$

This last expression contains the sum over all divisors of j, i.e. the *divisor function* [Kr81, Bu88]

$$\sigma(j) = \sum_{r|j} r \ .$$

Sums over expressions $r|j$ play an important role in number theory [Kr81, Bu88]: Each mapping of the set of integers into the set of complex numbers are denoted as *number theoretical function.* For instance, $\tau(N)$ yields the number of all divisors of N, and $\sigma(N)$ gives the sum of all divisors of N. A number theoretical function f is *multiplicative* if $f(N_1 \cdot N_2) = f(N_1) \cdot f(N_2)$.

The sum of two functions f and g and the multiplication of a function f by a complex number α are defined by

$$(f+g)(N) = f(N) + g(N) \ ,$$

$$(\alpha f)(N) = \alpha f(N) \ .$$

But very interesting is the fact that there is a *number theoretical convolution* $f*g$ which is defined by

$$(f*g)(N) = \sum_{d|N} f(N/d) \cdot g(d)$$

where the sum is taken over all positive divisors of N including also 1 and N. This convolution of two number theoretical functions is a commutative and associative algebraical operation.

Now note that $c(m_t) = c_t$ because for any $t=1,2,3,...$ in the constructed convex polygons no segment within the sequence of lines may be refined, i.e. no line can be replaced by an enlarged number of lines without destruction of the convexity of the corresponding polygon. On the other side $m(c_t) = m_t$ holds.

In [Vo82], approximation formulas are given based on the theory developed here:

<u>Formula 4.3.-1:</u>

$$m_{approx}(c) = 0.1507\, c^{3/2} + 0.1 + \frac{2}{c^{1/2}} \ ,$$

$$c_{approx}(m) = 3.53\, (m-1)^{2/3} \ .$$

Here, $m_i(c)$ is the minimal side length of a square covering a convex polygon with c corners, and $c_a(m)$ is the maximal number of corners of a convex polygon included into a square with side length m.

The table 4.3.-2 contains some examples for these formulas with respect to the exact values given in table 4.3.-1. Finally, the maximal possible corner number of convex polygons in $2^k \cdot 2^k$-images (which are of interest in image processing) shall be given in table 4.3.-3.

Table 4.3.-2: Approximative values of $m_i(c)$ and $c_a(m)$

c or m	$m_i(c)$	$m_{approx}(c)$	$c_a(m)$	$c_{approx}(m)$
1	1	2.251	1	0.000
2	2	1.940	4	3.530
3	2	2.038	6	5.604
4	2	2.306	8	7.343
5	3	2.679	9	8.895
6	3	3.131	10	10.322
7	4	3.647	12	11.656
8	4	4.217	13	12.917
9	5	4.836	14	14.120
10	6	5.498	16	15.273
15	10	9.371		20.505
18		12.080	24	23.339
20	14	14.026		25.135
33		29.016	36	35.580
40	38	38.541		40.596
92		133.291	72	71.417
128	218	218.514		89.189
178		358.135	112	111.282
218		485.299		127.472

Table 4.3.-3: Maximal corner number in $2^k \cdot 2^k$-images

k	0	1	2	3	4	5	6	7	8	9
m	1	2	4	8	16	32	64	128	256	512
c_{max}	1	4	8	13	22	35	56	90	142	226

4.3.3 Determination of Convex Hull

It is a well-known fact that the *convex hull* of a set of N disordered points in the real plane E^2 can be determined with $O(N \cdot \log N)$ operations [Pr85, Da88]. However the situation changes completely if the N points build up a connected object in the discrete space Z^2. In this case, the convex hull can be computed with $O(N)$ operations [Ki80, Hü81].

It was assumed in [Hü81] that a sequence $[p_1,p_2,\ldots,p_N]$ of points in the real plane E^2 is given whereby the straight line segments (p_1,p_2), $(p_2,p_3),\ldots,$ (p_{N-1},p_N), (p_N,p_1) represent in this order a crossing-free sequence (for instance, such a sequence results by contur following). The problem of N disordered points in Z^2 restricted in a $D \cdot D$-square was reduced to the problem of N ordered points with $O(D)$ operations. Therefore, the convex hull of any N points $p_i \epsilon Z^2$ can be determined with computational complexity $O(\max(N,D))$.

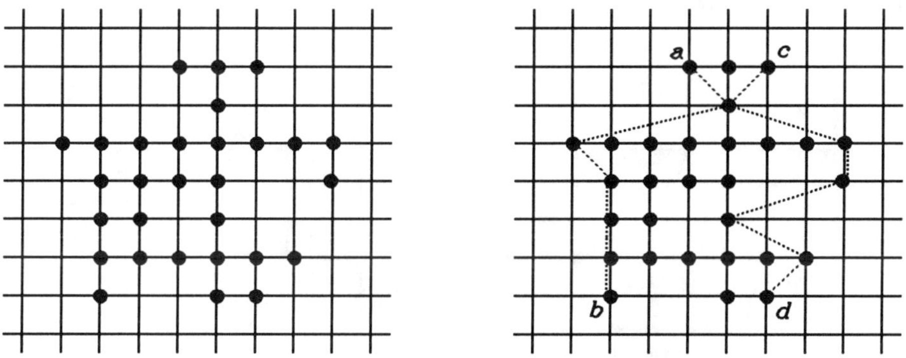

Fig. 4.3.-5: Connected point set (left) and line-extremal points (right)

We will here replace the assumption of a restricted point set by the assumption of connectedness of points $p_1,\ldots,p_N$. This assumption is very natural in digital image processing because we wish to determine mostly the convex hull of components obtained by graph search (section 1.3.3), i.e. the convex hull of connected point sets.

In figure 4.3.-5, a connected object is represented (left side). In each line y=const, the extremal points (i.e. points with minimal or maximal values of x-coordinate) will be determined (right side). This algorithmic step uses $O(N)$ operations:

```
0 → YMAX
MAX → YMIN
for I=1 to N do
begin
```

```
        min(Y_I,YMIN) → YMIN
        max(Y_I,YMAX) → YMAX
end
for Y=YMIN to YMAX do
begin
        0 → LISTMAX(Y)
        MAX → LISTMIN(Y)
end
for I=1 to N do
begin
        min(X_I,LISTMIN(Y_I)) → LISTMIN(Y_I)
        max(x_I,LISTMAX(Y_I)) → LISTMAX(Y_I)
end
```

Because the object is assumed as connected, it is always guaranteed that $\max(YMAX-YMIN+1,N)=N$, i.e. that we use $O(N)$ operations to build up the both lists of line-extremal points.

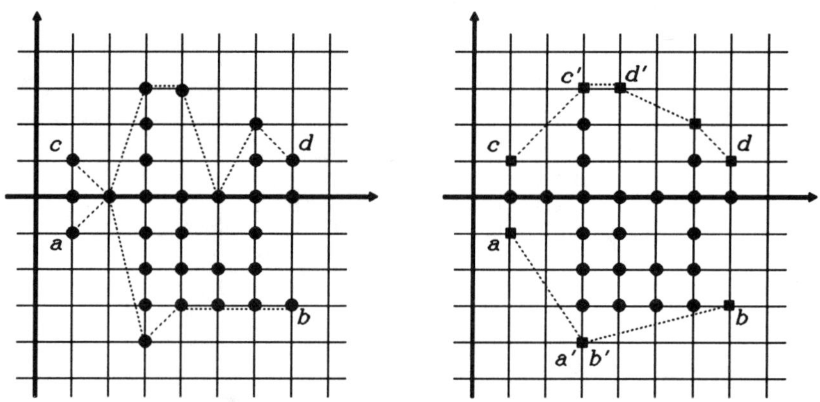

Fig. 4.3.-6: Determination of the convex hull of two functions

Now we have found two functions $x=listmin(y)$ and $x=listmax(y)$ for which convex hull functions are to be determined (figure 4.3.-6 left). The points *a,c, b,d, a',b'*, and *c',d'* are always (perhaps identical) points of the convex hull, and all the points $(y,listmin(y))$ and $(y,listmax(y))$ are ordered in respect of increasing values of y.

In the second algorithmic step, we determine now slope-monotonous increasing functions $x=f_{inc}(y)$ in the intervalls *c...c'* and *b'...b* and slope-monotonous decreasing functions $x=f_{dec}(y)$ in the intervalls *d'...d* and *a...a'* (figure 4.3.-6 right). Finally, the corner points of these slope-monotonous functions are the (ordered) corner points of the convex hull which were to be determined. Because this second algorithmic step needs also only $O(N)$ operations, the overall complexity for the convex-hull algorithm is $O(N)$.

Besides the importance of convex-hull algorithms, this example shows that structure and *complexity* of algorithms in computational geometry of Z^2 can be different drastically in comparison to corresponding algorithms for E^2.

4.3.4 Convexity in Z^n

Because we take the Z^n as embedded in the E^n, the notion of convexity has to be consistent with respect to both space types. Therefore, we use the following definition [Le80].

Def. 4.3.-4: A subset $C \subset Z^n$ is denoted as *convex* if and only if C contains all points convex-dependent on C. A point $p \in Z^n$ is *convex-dependent on C* if p can be represented by

$$p = \sum_{l=1}^{k} \lambda_l p_l \quad , \quad \lambda_l \geq 0 \quad , \quad \sum_{l=1}^{k} \lambda_l = 1 \quad , \quad p_l \in C \ .$$

For two dimensions, this definition agrees completely with definition 4.3.-2 because it is

$$x = (1-\lambda-\mu)\, x_1 + \lambda x_2 + \mu x_3 = \lambda_1 x_1 + \lambda_2 x_2 + \lambda_3 x_3 \ ,$$

$$y = (1-\lambda-\mu)\, y_1 + \lambda y_2 + \mu y_3 = \lambda_1 y_1 + \lambda_2 y_2 + \lambda_3 y_3 \ .$$

The advantage of definition 4.3.-2 is that the parameters λ and μ (and therefore also the parameters $\lambda_1,\lambda_2,\lambda_3$) can be determined immediatly whereas such a determination is not so simple if we use definition 4.3.-4.

For Euclidean spaces, definition 4.3.-4 is equivalent to the common definition of *line-convexity* that C is convex if, with two points $p_1,p_2 \in C$, all points of the straight line segment between p_1 and p_2 are points of C, too. But this equivalence is not valid in Z^n (figure 4.3.-7). A point set which is convex with respect to definition 4.3.-4 shall be denoted as *full-convex*. The set of figure 4.3-7 is line-convex but not full-convex. Therefore we have to work with definition 4.3.-4 if we wish to prove theorems for convex sets in Z^n.

For instance, there is the following theorem: Let be $C_1,C_2 \subset Z^n$ two convex point sets. Then the intersection $C=C_1 \cap C_2$ is also a convex set.

In the case of E^n, the common proof of this theorem concludes from $p_1,p_2 \in C$ that $p_1,p_2 \in C_1$ and $p_1,p_2 \in C_2$. We obtain $p \in C_1$ and $p \in C_2$ for all points p between p_1 and p_2. Therefore $p \in C=C_1 \cap C_2$ follows for all points p between p_1 and p_2, and C is convex, too.

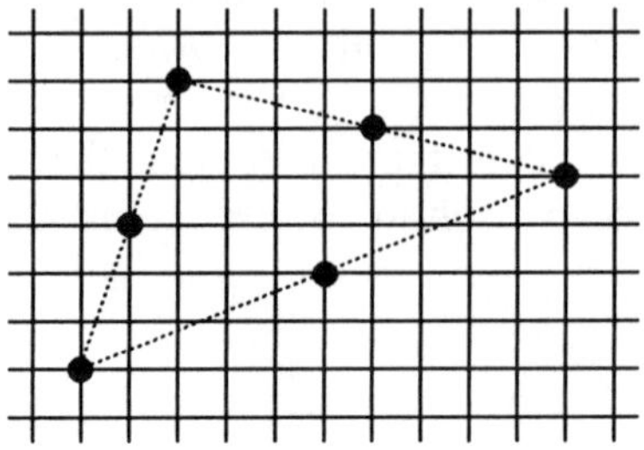

Fig. 4.3.-7: A line-convex point set which is not full-convex

This line-convex oriented argument is not sufficient to prove *full-convexity* of point sets because the proof fails for such points which don't lie "between" two other points (figure 4.3.-7). But if we consider the three corner points p_1,p_2,p_3 of the triangle in figure 4.3.-7 as points of intersection C and therefore as points of convex sets C_1 and C_2, then all points p of the triangle are triangular to p_1,p_2, p_3 in sense of definition 4.3.-3, and C is also convex, i.e. more exactly full-convex.

For image processing in Z^n, the construction of the convex hull and the proof of convexity of a finite set of lattice points are of more interest than the general mathematical investigation of line-convex or full-convex point sets. But contrary to $n=2$, the corresponding algorithms in three- and higher-dimensional orthogonal lattices are relatively cumbersome and complicated [Pr77, Da90]. A comprehensive bibliography of more than 370 papers on convexity problems and convexity algorithms was given by Ronse [Ro89a].

4.4 Approximative Motions

4.4.1 Pythagorean Rotations

In lattices, there are only a few motions in the strict sense, namely translations and tiltings (see section 3.2.1). But the connections to Euclidean geometry, caused and required by practical applications, open investigations into how we can described approximative motions [Gr86]. An *approximative motion* is defined by the expression

$$\begin{pmatrix} x_1' \\ x_2' \\ \vdots \\ x_n' \end{pmatrix} = \begin{pmatrix} ent(z_1) \\ ent(z_2) \\ \vdots \\ ent(z_n) \end{pmatrix} \quad , \quad \begin{pmatrix} z_1 \\ z_2 \\ \vdots \\ z_n \end{pmatrix} = \begin{pmatrix} r_{11} & r_{12} & \cdots & r_{1n} \\ r_{21} & r_{22} & \cdots & r_{2n} \\ \vdots & \vdots & \vdots & \vdots \\ r_{n1} & r_{n2} & \cdots & r_{nn} \end{pmatrix} \begin{pmatrix} x_1 \\ x_2 \\ \vdots \\ x_n \end{pmatrix} + \begin{pmatrix} s_1 \\ s_2 \\ \vdots \\ s_n \end{pmatrix}$$

where x_i and x_i' are integer coordinates, r_{ij} and s_i are real numbers, and *ent*(*z*) means the integer part of the real number *z*.

Such approximative motions *m* are frequently used in image processing. We wish to obtain a new image *I'* from an original image *I*, i.e. gray values $g_{I'}(x') = g_I(m^{-1}(x'))$. For each point $x' \in I'$ of the new image *I'*, we have to determine the corresponding point $x = m^{-1}(x')$ in the original image *I*, and the gray value $g_I(x)$ is to transfer to the point *x*'. Therefore, properly speaking, we are interested in the inverse transformation m^{-1} instead in the transformation *m*.

Because of the non-linearity of the above equations caused by the function *ent*, we cannot invert them so that we have to invert the non-digitized equations before we determine the integer coordinates of point *x*.

We will investigate here rational rotations in Z^2 where some selected points of the new image will be mapped exactly into lattice points of the original image. In figure 4.4.-1, such a rotation is represented. The side length $c = AB$ of the square is 5, and we can obtain *c* also by the side *A'B'* of the rotated square because the equation $5^2 = 3^2 + 4^2$ is fulfilled.

Def. 4.4.-1: Given a square with corners at lattice points in Z^2. If the side length *c* is an integer then we call the square a *Pythagorean square*.

In number theory, it is shown that all integer solutions of $a^2 + b^2 = c^2$ with $a,b,c > 0$, $gcd(a,b,c) = 1$ and even *b* are given by

$$a = u^2 - v^2 \quad , \quad b = 2uv \quad , \quad c = u^2 + v^2$$

if $u > v > 0$, $gcd(u,v) = 1$ and $u \not\equiv v$ *modulo* 2. There are inifinitely many of such number tripels [Kr81]. Some of Pythagorean tripels are given in table 4.4.-1.

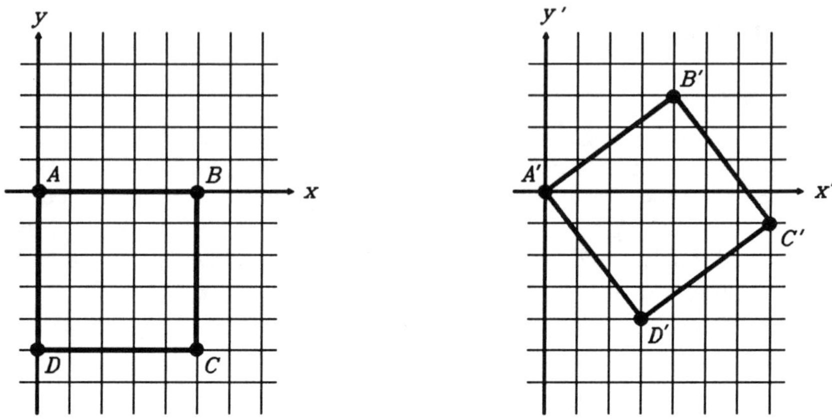

Fig. 4.4.-1: Rotation of a Pythagorean square

Table 4.4.-1: Pythagorean tripels a,b,c with $a^2+b^2=c^2$

u	v	a u^2-v^2	b $2uv$	c u^2+v^2	atan(min(a,b)/max(a,b))
2	1	3	4	5	36.87 ⁰
3	2	5	12	13	22.62 ⁰
4	1	15	8	17	28.07 ⁰
4	3	7	24	25	16.26 ⁰
5	2	21	20	29	43.60 ⁰
5	4	9	40	41	12.68 ⁰
6	1	35	12	37	18.93 ⁰
6	5	11	60	61	10.39 ⁰
7	2	45	28	53	31.89 ⁰
7	4	33	56	65	30.51 ⁰
7	6	13	84	85	8.80 ⁰

It is possible to approximate the value of tan(α) for any given rotation angle α with sufficient precision (of course, then the value of c will become greater and greater). But for any tripel (a,b,c) of table 4.4.-1, a super lattice of squares of the new image I' (left side of figure 4.4.-2) can be determined by a super-lattice of rotated squares of original image I (right side of figure 4.4.-1).

Suppose that we have determined the coordinates x,y of the original points for all points of a basic square of the new image I', the coordinates x,y of any other points (x',y') can be obtained by

$$\xi' \equiv x' \ modulo \ c \quad , \quad \xi = ent\left(m_x^{-1}(\xi')\right) ,$$
$$\eta' \equiv y' \ modulo \ c \quad , \quad \eta = ent\left(m_y^{-1}(\eta')\right) ,$$

$$\begin{pmatrix} x \\ y \end{pmatrix} = \begin{pmatrix} a & -b \\ b & a \end{pmatrix} \begin{pmatrix} ent(x'/c) \\ ent(y'/c) \end{pmatrix} + \begin{pmatrix} \xi \\ \eta \end{pmatrix} .$$

The inverse transformations m^{-1} have to be determined only for c^2 points. For all other points, we have only to calculate the offsets of the rotated squares and to read the difference coordinates ξ,η from a look-up-table. But this method is only practical if the squares are small.

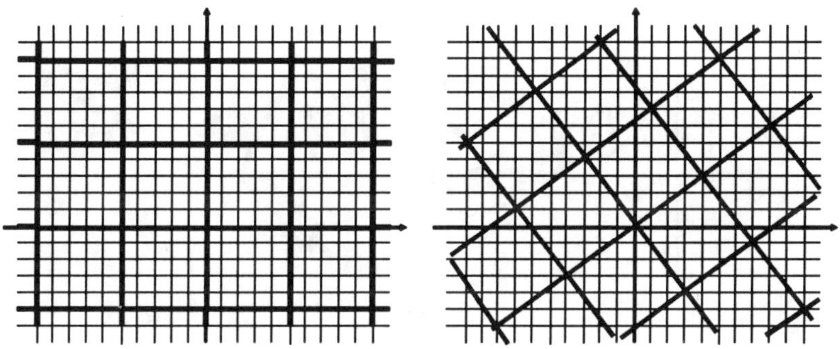

Fig. 4.4.-2: Pythagorean rotation of the super lattice

4.4.2 Shear Transformations

In this section, we will investigate a method to realize approximate rotations which can be supported by hardware. Up to now, only translations, rotations (tiltings), and reflections has been discussed. But there is yet another group of transformations, namely *shear transformations*. A two-dimensional shear transformation is given by

$$\begin{pmatrix} x' \\ y' \end{pmatrix} = \begin{pmatrix} 1 & \alpha \\ \beta & 1 \end{pmatrix} \begin{pmatrix} x \\ y \end{pmatrix}$$

where $\alpha=0$ or $\beta=0$. Such a transformation shifts only rows defined by y=const with $\Delta x=\alpha y$ or shifts only columns defined by x=const with $\Delta y=\beta x$. Of course, this transformation is not a motion because neither distances nor angles remain unchanged.

But each rotation can be represented by three consecutive shear transformations [We76, Ta86, Sh90]:

Formula 4.4.-1:

$$\begin{pmatrix} \cos\varphi & \sin\varphi \\ -\sin\varphi & \cos\varphi \end{pmatrix} = \begin{pmatrix} 1 & 0 \\ -\tan(\varphi/2) & 1 \end{pmatrix} \begin{pmatrix} 1 & \sin\varphi \\ 0 & 1 \end{pmatrix} \begin{pmatrix} 1 & 0 \\ -\tan(\varphi/2) & 1 \end{pmatrix}.$$

There is a degeneration for $\varphi/2=\pm 90^0$. In this case, we have to perform a 180^0-rotation, i.e. a simple tilting with $x'=-x$ and $y'=-y$. In figure 4.4.-3, a 70^0-rotation is realized by a first shearing in y-direction, a second shearing in x-direction, and a third shearing once more in y-direction.

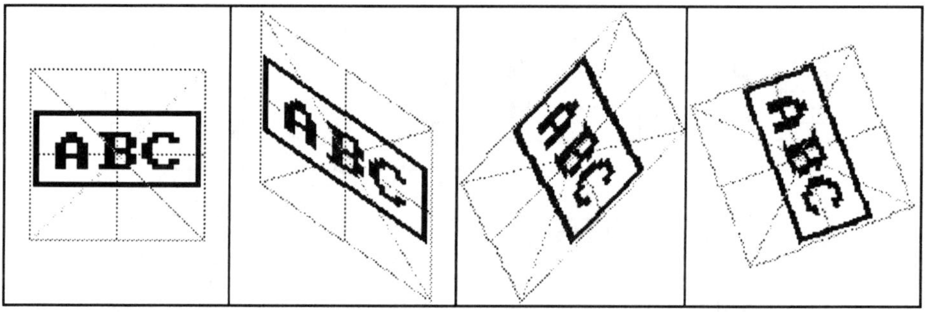

Fig. 4.4.-3: 70^0-rotation realized by three consecutive shear transformations

Each single shear transformation needs only a constant integer shift $\Delta x=ent(s\cdot y)$ of all points of a row y=const along the row or a shift $\Delta y=ent(s\cdot x)$ of all points of a column x=const along the column whereby s is any real number. Because of that, fast hardware realizations or fast software realizations are possible.

The equations $\Delta x=ent(s\cdot y)$ and $\Delta y=ent(s\cdot x)$ are equations of digital straight lines (see section 4.2.2). Therefore, we will describe any rotation nearly by approximated values for $\sin\varphi$ and $\tan(\varphi/2)$ so that periodicities arise in the sequence of shift values Δx and Δy.

In Z^n, a general shear transformation is defined by an $n\cdot n$-matrix with coefficients $a_{ii}=1$ and $a_{ij}=0$ for all $i>j$. It is possible to replace n-dimensional rotations by consecutive shear transformations [Ca80].

Especially in the three-dimensional Euclidean space, we can describe general rotations by *Eulerian angles*. Let

φ be the rotation angle in respect of the x-axis,
ψ be the rotation angle in respect of the y-axis,
χ be the rotation angle in respect of the z-axis, (see figure 4.4.-4).

These three rotations can be described by the matrices $\boldsymbol{R}_x$, $\boldsymbol{R}_y$, and $\boldsymbol{R}_z$:

$$\boldsymbol{R}_x = \begin{pmatrix} 1 & 0 & 0 \\ 0 & \cos\varphi & -\sin\varphi \\ 0 & \sin\varphi & \cos\varphi \end{pmatrix}, \quad \boldsymbol{R}_y = \begin{pmatrix} \cos\psi & 0 & \sin\psi \\ 0 & 1 & 0 \\ -\sin\psi & 0 & \cos\psi \end{pmatrix},$$

$$\boldsymbol{R}_z = \begin{pmatrix} \cos\chi & -\sin\chi & 0 \\ \sin\chi & \cos\chi & 0 \\ 0 & 0 & 1 \end{pmatrix}.$$

It is possible to compose any general rotation from three single Eulerian rotations what can be done by multiplication of the matrices $\boldsymbol{R}_x$, $\boldsymbol{R}_y$, and $\boldsymbol{R}_z$. Because these matrix multiplications are non-commutative, the resulting transformation matrix is dependent from the order of the single matrices. For instance, we obtain $\boldsymbol{R}_x\boldsymbol{R}_y\boldsymbol{R}_z \neq \boldsymbol{R}_z\boldsymbol{R}_y\boldsymbol{R}_x$.

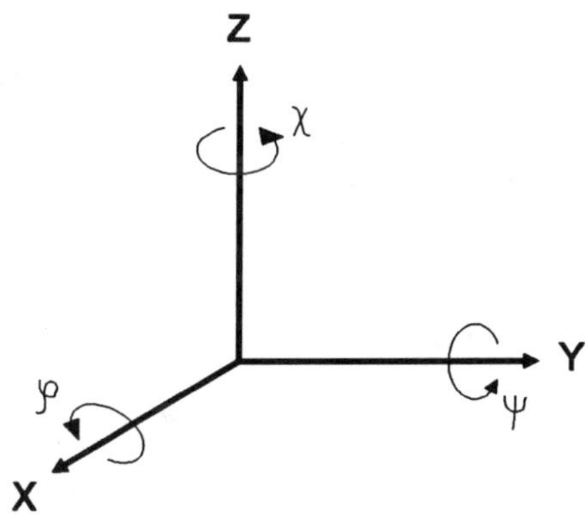

Fig. 4.4.-4: Eulerian angles for three-dimensional rotations

If we perform at first a rotation in respect of the z-axis (multiplication with $\boldsymbol{R}_z$), then a rotation in respect of the y-axis, and finally a rotation in respect of the x-axis, then the resulting orthogonal transformation matrix $\boldsymbol{R} = \boldsymbol{R}_x\boldsymbol{R}_y\boldsymbol{R}_z$ is given by

$$\begin{pmatrix} \cos\psi\cos\chi & -\cos\psi\sin\chi & \sin\psi \\ \cos\varphi\sin\chi+\sin\varphi\sin\psi\cos\chi & \cos\varphi\cos\chi-\sin\varphi\sin\psi\sin\chi & -\sin\varphi\cos\psi \\ \sin\varphi\sin\chi-\cos\varphi\sin\psi\cos\chi & \sin\varphi\cos\chi+\cos\varphi\sin\psi\sin\chi & \cos\varphi\cos\psi \end{pmatrix}.$$

Using the above formula for replacement of a two-dimensional rotation by a product $A_\varphi B_\varphi A_\varphi$ of shear matrices, it is

$$\boldsymbol{R} = A_\varphi^{(x)} B_\varphi^{(x)} A_\varphi^{(x)} A_\psi^{(y)} B_\psi^{(y)} A_\psi^{(y)} A_\chi^{(z)} B_\chi^{(z)} A_\chi^{(z)} \quad .$$

4.4.3 General Affine Transformations

Finally, we will investigate general *affine transformations* of E^n and one-to-one mappings of Z^n which approximate any given equivolume affine transformations. The equivolume property is necessary because whether no new lattice points are permitted (as it would be the case for volume enlargement) nor lattice points are to be neglected.

Equivolume affine transformations described by a transformation matrix $\boldsymbol{A}$ includes translations, reflections, rotations, and shear transformations. The are characterized by $|\boldsymbol{A}| = 1$.

Besides translations, a general affine transformation $\boldsymbol{A}$ can be decomposed in a product of special affine transformations [Sh90]:

$$\boldsymbol{A} = \boldsymbol{TRSF} \quad .$$

Here, $\boldsymbol{T}$ is a generalized tilting matrix, $\boldsymbol{R}$ is an orthogonal rotation matrix, $\boldsymbol{S}$ is the matrix of a shear transformation, and $\boldsymbol{F}$ is a matrix which describes factor scaling of single coordinates:

$$\boldsymbol{T}: \quad t_{ii} \in \{+1, -1\} \ , \ a_{ij} = 0 \quad \text{for } i \neq j \ , \ |\boldsymbol{T}| = -1 \ ,$$

$$\boldsymbol{R}: \quad \boldsymbol{R}^T\boldsymbol{R} = \boldsymbol{I} \ , \ |\boldsymbol{R}| = +1 \ ,$$

$$\boldsymbol{S}: \quad s_{ii} = +1 \ , \ s_{ij} = 0 \quad \text{for } i > j \ ,$$

$$\boldsymbol{F}: \quad f_{ii} > 0 \ , \ f_{ij} = 0 \quad \text{for } i \neq j \ , \ |\boldsymbol{F}| = +1 \ .$$

Shizawa gives a theorem and a corresponding algorithm, that each equivolume affine transformation of the n-dimensional Euclidean space E^n is decomposable into n^2-1 fundamental shear transformations (he speaks of *skew transformations*) and some fundamental tilting transformations (he speaks of reflective transformations).

It is essential to emphasize that the method of shear transformation guarantees equivolume one-to-one mappings of Z^n. The combination of such transformations is also a one-to-one mapping, i.e. an invertible mapping.

5 Discrete Functions

5.1 One-Dimensional Periodical Discrete Functions

5.1.1 Functions

Let D be any set of elements and let R be any other set ($D=R$ is allowed). If we have a certain law L according to which to each element $d \in D$ an element $r \in R$ is associated, we say that this law is a *function* defined on D.

We symbolize that association by $r=L(d)$. The set D is the *domain of definition* of the function, and the set $R' \subseteq R$ is the *range of values* where R' consists of all that elements $r \in R$ for which there is a $d \in D$ with $L(d)=r$.

For two functions F and G defined on the same set D, we can define the sum, the difference, the product, and the quotient by

$$H = F + G \quad , \quad H(d) = F(d) + G(d) \quad , \quad d \in D \ ,$$

$$H = F - G \quad , \quad H(d) = F(d) - G(d) \quad , \quad d \in D \ ,$$

$$H = F \cdot G \quad , \quad H(d) = F(d) \cdot G(d) \quad , \quad d \in D \ ,$$

$$H = \frac{F}{G} \quad , \quad H(d) = \frac{F(d)}{G(d)} \quad , \quad d \in D \ .$$

However, it is not necessary that all these operations are defined for the elements of R (see section 5.2.1).

In school mathematics, it is assumed that the domain D and the range R' are equal to subsets of real numbers. But in general, we cannot assume this condition. For instance, the domain of definition can be the set P of all points of a neighborhood structure $[P,N]$ or the set of all points of an n-dimensional incidence structure.

The range R' is in image processing often given by the set $R'=\{0,1,\ldots,255\}$ of gray values. It is clear that the addition of two image functions F and G cannot be defined by a simple numerical addition. We have to use laws of addition as for example

$$H = F \oplus G \quad , \quad H(d) = \min(F(d) + G(d) \, , 255) \quad , \quad d \in D \ ,$$

$$H = F \oplus G \quad , \quad H(d) = ent\left(\frac{F(d) + G(d)}{2} \right) \quad , \quad d \in D \ ,$$

$$H = F \oplus G \quad , \quad H(d) = ent\left(\frac{F(d) + G(d)}{1 + \dfrac{F(d) \cdot G(d)}{255^2}} \right) \quad , \quad d \in D \ .$$

The third of these addition laws is a "relativistic" law because 255 is the upper limit of all possible gray values so that $255 \oplus r = 255$.

Here, the function $g = ent(x)$ means the greatest integer g which is smaller than or equal to x ("*entier*"), i.e. $g \leq x < g+1$. For this function, there are the Gaussian formula

$$\sum_{k=1}^{l} ent(kx) + \sum_{k=1}^{ent(lx)} ent\left(\frac{k}{x}\right) = l \cdot ent(lx)$$

for non-integer values x and the Hermitean formula

$$\sum_{k=0}^{l-1} ent\left(x + \frac{k}{l}\right) = ent(lx)$$

for any real x.

We use here a one-dimensional closed neighborhood structure with N points as domain of definition. In figure 5.1.-1, such a structure is represented which is so to speak a closed one-dimensional gray value image. The number N is thereby the characteristical quantity of the structure (for higher-dimensional domains of definition, closed toroidal structures are likewise the suitable supports).

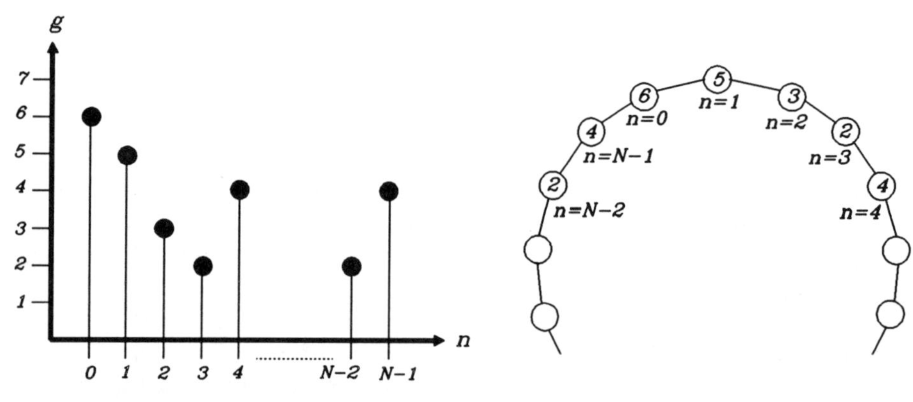

Fig. 5.1.-1: Structure of a one-dimensional periodical gray value image

Each point p_n of this finite neighborhood structure has two neighbors p_{n-1} and p_{n+1}. We can attach the points p_n to the integer numbers $n \in \{0,1,\ldots,N-1\}$ whereby we have to guarantee $p_{-1} = p_{N-1}$ and $p_N = p_0$, i.e. in general

$$p_{n \pm N} = p_n \ .$$

This condition requires also that $n \pm N = n$, which is possible only then when the quantities n are not common integers but representatives for number theoretical *residue classes*. Therefore, we write the relationship of the above equation in numerical form as

$$n \pm k \cdot N \equiv n \quad modulo \ N$$

which means that we are interested not in the common value of the integers n but in the residue of n which we obtain by division by N.

5.1.2 Space of Periodical Discrete Functions

Now, let R be the set of values to which the points p_n of a one-dimensional finite incidence structure are mapped by a function F (we suppose in the following that R is a commutative field, for instance the set of all real numbers). We write $F(p_n) = F_n \in R$ and call this mapping a *discrete periodical function*.

Such a function can be described unambiguously by an N-tupel $(F_0, \ldots, F_{N-1})$ of elements F_n of set R. Instead of a discrete function, we can speak also of a N-dimensional *vector* for whose coefficients $F_{n \pm N} = F_n$ is a necessary condition.

Def. 5.1.-1: The set $\boldsymbol{F}$ of discrete periodical functions $F \in \boldsymbol{F}$ with coefficients $F_n \in R$ out of a set R is a *function space* with following properties:

- There is exactly one element $H = F + G$ as *sum* of the two functions $F, G \in \boldsymbol{F}$ which is defined by the addition of the coefficients $H_n = F_n + G_n$, $n \in \{0, \ldots, N-1\}$.
- There exists an associative law $(F+G)+H = F+(G+H)$, and there is a commutative law $F+G = G+F$.
- There is a null element $O = (0, \ldots, 0)$ with $F+O = F$ for all $F \in \boldsymbol{F}$, and there is an inverse element $-F = (-F_0, \ldots, -F_{N-1})$ for all $F \in \boldsymbol{F}$ so that $F+(-F) = O$.
- A multiplication is defined between elements $r \in R$ and elements $F \in \boldsymbol{F}$ so that $G = r \cdot F = (r \cdot F_0, \ldots, r \cdot F_{N-1})$.

These properties and requirements can be summarized in the statement that $\boldsymbol{F}$ is an *algebra over R* (see section 5.2.2). We can define a *pseudo scalar product* $\langle F, G \rangle$ between two periodical functions F and G which is an element of R:

$$\langle F \, , G \rangle = \sum_{n=0}^{N-1} F_n \, G_n \ .$$

It can be easily proved that this pseudo scalar product (shortly designed as *P-product*) is characterized by the following properties:

$$\langle rF, G\rangle = r\langle F, G\rangle \quad \text{with} \quad r \in R,$$

$$\langle F, G\rangle = \langle G, F\rangle,$$

$$\langle F+G, H\rangle = \langle F, H\rangle + \langle G, H\rangle .$$

The function space $\boldsymbol{F}$ is an Euclidean vector space and the expression $<F,G>$ is a *scalar product* if the condition $<F,F>\geq 0$ is fulfilled for any $F \epsilon \boldsymbol{F}$.

There are some further notions which we will use in the following developments. First, we have the already mentioned *null function* $\boldsymbol{O}$ as null element. At second, we introduce N *basic functions*

$$E^{(n)} = (\delta_{0-n}, \delta_{1-n}, \ldots, \delta_{(N-1)-n}) \quad , \quad n \in \{0,\ldots,N-1\}$$

with

$$\delta_k = \begin{cases} 1 & \text{for} \quad k=0 \\ 0 & \text{else} \end{cases}$$

as *Kronecker symbol.*

These basic functions have zero coefficients up to the n-th coefficient which has the value 1, i.e.

$$E_n^{(k)} = \delta_{n-k} .$$

Using the basic functions $E^{(0)}, E^{(1)},\ldots,E^{(N-1)}$ which are a basis of the vector space $\boldsymbol{F}$, we can write every function $F \epsilon \boldsymbol{F}$ as a weighted sum over the $E^{(n)}$:

$$\begin{aligned} F &= (F_0, F_1, \ldots, F_{N-1}) \\ &= F_0(1,0,0\ldots,0) + \\ &\qquad + F_1(0,1,0,\ldots,0) + \ldots \\ &\qquad\qquad \ldots + F_{N-1}(0,0,0,\ldots,1) \\ &= \sum_{n=0}^{N-1} F_n E^{(n)} . \end{aligned}$$

Furthermore, the function $F^{\leftrightarrow}$ shall be the *reflected function* of F with the notation $F_n^{\leftrightarrow} = F_{-n}$. A *symmetrical function* F is characterized by $F^{\leftrightarrow}=F$, and for an *anti-symmetrical function* $F^{\leftrightarrow}=-F$ must be fulfilled. Each function F can be decomposed in a sum of a symmetrical and an anti-symmetrical function:

$$F = \frac{F + F^{\leftrightarrow}}{2} + \frac{F - F^{\leftrightarrow}}{2} .$$

5.1.3 LSI-Operators and Convolutions

For a *linear operator* $\boldsymbol{L}$ which transforms functions $F \epsilon \boldsymbol{F}$ in new functions $G = \boldsymbol{L}F$, we have for $F,G \epsilon \boldsymbol{F}$ and $a,b \epsilon R$

$$\boldsymbol{L}(a \cdot F + b \cdot G) = \boldsymbol{L}(a \cdot F) + \boldsymbol{L}(b \cdot G) = a \cdot \boldsymbol{L}F + b \cdot \boldsymbol{L}G$$

and therefore with the basic functions $E^{(n)}$

$$\boldsymbol{L}F = \boldsymbol{L}\left(\sum_{n=0}^{N-1} F_n E^{(n)}\right) = \sum_{n=0}^{N-1} F_n \boldsymbol{L}(E^{(n)}) .$$

The functions $\boldsymbol{L}(E^{(n)})$ can be written as

$$\boldsymbol{L}E^{(n)} = \sum_{m=0}^{N-1} L_m^{(n)} E^{(m)}$$

because they can be represented as a sum over basic functions, too. Therefore, the application of the linear operator $\boldsymbol{L}$ to a function F is equivalent to a multiplication of the function represented by a vector $F=(F_0,\ldots,F_{N-1})^T$ using a matrix L whose coefficients are given by the above formula (the superscipt T means the *transposition* of the vector):

$$\begin{pmatrix} G_0 \\ G_1 \\ \cdot \\ \cdot \\ G_{N-1} \end{pmatrix} = \begin{pmatrix} L_0^{(0)} & L_0^{(1)} & \cdots & L_0^{(N-1)} \\ L_1^{(0)} & L_1^{(1)} & \cdots & L_1^{(N-1)} \\ \cdot & \cdot & \cdots & \cdot \\ \cdot & \cdot & \cdots & \cdot \\ L_{N-1}^{(0)} & L_{N-1}^{(1)} & \cdots & L_{N-1}^{(N-1)} \end{pmatrix} \begin{pmatrix} F_0 \\ F_1 \\ \cdot \\ \cdot \\ F_{N-1} \end{pmatrix} .$$

A very important operation in image processing is the *convolution* $F*G$ between two functions $F,G \epsilon \boldsymbol{F}$. For two one-dimensional discrete periodical functions, the n-th coefficient $H_n=(F*G)_n$ of the convolution $H=F*G$ is defined by

$$(F*G)_n = \sum_{k=0}^{N-1} F_{n-k} G_k .$$

The convolution of two functions is a distributive, commutative, and associative operation:

$$F*(G+H) = F*G + F*H \ ,$$

$$F*G = G*F \ ,$$

$$F*(G*H) = (F*G)*H \ .$$

An operator $\boldsymbol{C_F}$ is called *convolution operator*, if a discrete periodical function F exists such that

$$\boldsymbol{C_F} G = F*G$$

for all functions $G \epsilon \boldsymbol{F}$. Because of that, each convolution operator is a linear operator.

The *k-shift* of a function G yields a function $G^{|k}$ whose coefficients are defined by the formula

$$G_n^{|k} = G_{n-k} \ .$$

Using this definition, we obtain for the basic functions the relationship

$$E^{(k)} = (E^{(0)})^{|k} \ .$$

An operator $\boldsymbol{L}$ is called *shift-invariant*, if

$$\boldsymbol{L} G^{|k} = (\boldsymbol{L} G)^{|k} \ ,$$

i.e. if the application of $\boldsymbol{L}$ to the shifted function $G^{|k}$ can be obtained also by the shift of the result of $\boldsymbol{L}G$. Each convolution operator $\boldsymbol{C_F}$ described by a function F is shift-invariant, because we can prove that

$$\begin{aligned}
(\boldsymbol{C_F} G^{|k})_n &= (F*G^{|k})_n = \sum_{m=0}^{N-1} F_m G_{n-m}^{|k} \\
&= \sum_{m=0}^{N-1} F_m G_{n-k-m} = (F*G)_{n-k} \\
&= ((F*G)^{|k})_n = ((\boldsymbol{C_F} G)^{|k})_n \ .
\end{aligned}$$

Linear shift-invariant operators are denoted as *LSI-operators* so that we have shown that each convolution operator is an LSI-operator.

On the other side, let us assume, that $\boldsymbol{L}$ is an LSI-operator. Writing each function G as a sum

$$G = \sum_{k=0}^{N-1} G_k E^{(k)}$$

of basic-functions $E^{(k)}$ and using

$$(E^{(k)})_n = (E^{(0)})_{n-k} \ ,$$

we obtain for an LSI-operator $\boldsymbol{L}$

$$(\boldsymbol{L}G)_n = \sum_{k=0}^{N-1} G_k(\boldsymbol{L}E^{(k)})_n = \sum_{k=0}^{N-1} G_k((\boldsymbol{L}E^{(0)})\,|^k)_n$$

$$= \sum_{k=0}^{N-1} G_k(\boldsymbol{L}E^{(0)})_{n-k} = (G*\boldsymbol{L}E^{(0)})_n \ .$$

The function $\boldsymbol{L}E^{(0)}$ is denoted as *discrete point spread function* (PSF) because in image processing, LSI-operators are mostly given by locally restricted filters (see section 5.4.3).

We have proved the following theorem which shows the central importance of discrete convolution for image processing:

Theorem 5.1.-1: An operator in function space $\boldsymbol{F}$ is a linear and shift-invariant operator (a so-called *LSI-operator*), if and only if it is a convolution operator. The basic function $E^{(0)}$ is the unit-element for the convolution as "multiplicative" operation in $\boldsymbol{F}$.

Because there exist properties like associativity and commutativity of the operation $*$, the algebraic structure $[\boldsymbol{F},E^{(0)},*]$ is an *Abelian monoid*, i.e. a commutative semigroup with unit element (see section 5.2.2 for the exact algebraical definition, and see chapter 5.4 for the approximative inversion of convolution).

The convolution $F*G$ of two functions $F,G\in\boldsymbol{F}$ is equivalent to the application of an LSI-operator $\boldsymbol{C_F}$ to the function G, and it is equivalent to the multiplication of the vector $G=(G_0,G_1,\ldots,G_{N-1})$ with a (*circulant*) matrix C_F. This one-to-one correspondence is caused generally by the linear properties of convolution, and it can be represented for one-dimensional functions in the usual matrix form for one-dimensional functions:

$$\boldsymbol{L}G = \boldsymbol{C_F}G = C_F G = \begin{pmatrix} F_0 & F_{N-1} & F_{N-2} & . & . & . & F_2 & F_1 \\ F_1 & F_0 & F_{N-1} & . & . & . & F_3 & F_2 \\ F_2 & F_1 & F_0 & . & . & . & F_4 & F_3 \\ . & . & . & . & . & . & . & . \\ . & . & . & . & . & . & . & . \\ F_{N-2} & F_{N-3} & F_{N-4} & . & . & . & F_0 & F_{N-1} \\ F_{N-1} & F_{N-2} & F_{N-3} & . & . & . & F_1 & F_0 \end{pmatrix} \begin{pmatrix} G_0 \\ G_1 \\ G_2 \\ . \\ . \\ G_{N-2} \\ G_{N-1} \end{pmatrix} .$$

On the other hand, every matrix describes a linear operator, but any matrix is not necessary a convolution operator.

5.1.4 Products of Linear Operators

Like in sections 1.4.4/5, we will investigate here the relationships between the most important linear operators in function space. Up to now, we know the reflection operator $\boldsymbol{R}$ with $(\boldsymbol{R}F)_n=F_{-n}$ and the convolution operator $\boldsymbol{C}_F$, where $\boldsymbol{R}$ is a linear operator but not an LSI-operator).

The k-shift of a function can also be represented by an operator $\boldsymbol{S}^{|k}$ with

$$(\boldsymbol{S}^{|k}F)_n = F_{n-k} \ .$$

Further, we use the identy operator $\boldsymbol{I}$. With N as characteristic quantity of the investigated functions, the shift operators $\boldsymbol{S}^{|k}$ can be represented by a k-fold power of $\boldsymbol{S}=\boldsymbol{S}^{|1}$ so that $\boldsymbol{S}^{|N}=\boldsymbol{S}^N=\boldsymbol{I}$.

Finally, the reflection of a reflected function is equal to the original function. Therefore $\boldsymbol{RR}=\boldsymbol{R}^2=\boldsymbol{I}$. There exists an inverse operator $\boldsymbol{R}^{-1}=\boldsymbol{R}$ and an inverse operator $\boldsymbol{S}^{-1}=\boldsymbol{S}^{|-1}=\boldsymbol{S}^{N-1}$, so that we obtain $(\boldsymbol{S}^{|k})^{-1}=\boldsymbol{S}^{|-k}=(\boldsymbol{S}^{|-1})^k$.

For a given convolution equation $F*X=\boldsymbol{C}_FX=G$, a solution X only exists if the operator $\boldsymbol{C}_F$ is invertible, i.e. if the inverse matrix of the matrix $\boldsymbol{C}_F$ exists. This condition requires that the determinant of the matrix C_F is nonzero. It is not possible that $\det(C_F)\neq 0$ for any function F. For instance, a constant function F_n=const causes $det(C_F)=0$.

Taking into account that $\boldsymbol{R}^{-1}$ and $\boldsymbol{S}^{-1}$ exist, and that the convolution is commutative (i.e. $\boldsymbol{C}_F\boldsymbol{C}_G=\boldsymbol{C}_G\boldsymbol{C}_F$), the table 5.1.-1 summarizes all binary operator products. Thereby the operator products $\boldsymbol{AB}$ consist of operators $\boldsymbol{A}$ of the left column and of operators $\boldsymbol{B}$ of the upper row.

As example for the proof of the corresponding products in table 5.1.-1, we will prove the equation $\boldsymbol{RC}_F=\boldsymbol{C}_{F^{\leftrightarrow}}\boldsymbol{R}$. It is

$$(\boldsymbol{R}\boldsymbol{C}_F G)_n = (\boldsymbol{C}_F G)_{-n} = \sum_{k=0}^{N-1} F_{(-n)-k} G_k$$

$$= \sum_{k=0}^{N-1} F^{\leftrightarrow}_{n-(-k)} G^{\leftrightarrow}_{-k} = \sum_{l=0}^{N-1} F^{\leftrightarrow}_{n-l} G^{\leftrightarrow}_{l}$$

$$= (F^{\leftrightarrow} * G^{\leftrightarrow})_n = (\boldsymbol{C}_{F^{\leftrightarrow}}\boldsymbol{R}G)_n \ .$$

This equation can be written as $\boldsymbol{R}(F*G)=(\boldsymbol{R}F)*(\boldsymbol{R}G)$, i.e. the reflection of a convolution product of two functions is equal to the convolution product of the two reflected functions.

Table 5.1.-1: Operator products $\boldsymbol{AB}$ in function space

	$\boldsymbol{B}=\boldsymbol{I}$	$\boldsymbol{C}_F$	$\boldsymbol{R}$	$\boldsymbol{S}$
$\boldsymbol{A}=\boldsymbol{I}$	$\boldsymbol{I}$	$\boldsymbol{C}_F$	$\boldsymbol{R}$	$\boldsymbol{S}$
$\boldsymbol{C}_G$	$\boldsymbol{C}_G$	$\boldsymbol{C}_G\boldsymbol{C}_F$	$\boldsymbol{R}\boldsymbol{C}_{G^{\leftrightarrow}}$	$\boldsymbol{S}\boldsymbol{C}_G$
$\boldsymbol{R}$	$\boldsymbol{R}$	$\boldsymbol{C}_{F^{\leftrightarrow}}\boldsymbol{R}$	$\boldsymbol{I}$	$\boldsymbol{S}^{-1}\boldsymbol{R}$
$\boldsymbol{S}$	$\boldsymbol{S}$	$\boldsymbol{C}_F\boldsymbol{S}$	$\boldsymbol{R}\boldsymbol{S}^{-1}$	$\boldsymbol{S}\boldsymbol{S}$

Finally, an interesting relationship between any LSI-operators $\boldsymbol{L}$ and convolution products can be derived. Because of theorem 5.1.-1, an LSI-operator $\boldsymbol{L}$ corresponds to a convolution operator $\boldsymbol{C}_L$. Further, the convolution fulfills the commutative and the associative law. Therefore we obtain

$$\boldsymbol{L}(F*G) = (\boldsymbol{L}F)*G = F*(\boldsymbol{L}G)$$

for any LSI-operator $\boldsymbol{L}$.

5.2 Algebraic Theory of Discrete Functions

5.2.1 Domain of Definition and Range of Values

It is possible and necessary to develop a logically closed theory of discrete functions using algebraic methods. The classical mathematical tools of analog signal processing are analysis and calculus. However, there are some discrete algorithms in image processing having no correspondences in analog signal processing. Indeed, the fast Fourier transform algorithm (FFT) cannot be applied for computing Fourier integrals. The finite discrete convolution uses algebraic methods, and number theoretical transforms (NTT) have nothing to do with analysis, too.

This means that the theory of discrete functions or the methods of discrete signal processing cannot be subordinated to classical theory of processing of continuous signals. Consequently, the structural characteristics of discrete phenomenons must be the starting point to develop a theory of discrete signals.

First, we have to replace the set R of real numbers by the set Z of integers (or by other sets), and we have to replace analysis by algebra. Second, the finite character of all numerical algorithms suggests using finite structures as domains of definition as we have found such structures to be toroidally closed homogeneous incidence structures.

The basis of our investigations are translative neighborhood structures $[P,N]$. If we fix an arbitrary point $o \in P$ as origin, the mapping $\boldsymbol{T} \to P$, $p = T(o)$ is a one-to-one mapping from the group $\boldsymbol{T}$ of all translations onto the set P.

Denoting the translation $T(o)=p$ by T_p, we induce in P an additive group structure $p+q$ which is defined by

$$T_{p+q} = T_p T_q \quad \text{with} \quad p, q \in P \ .$$

Particularly, $T_o = I$ is the identical translation, T_p^{-1} corresponds to $-p$, and $T_p T_q^{-1}$ corresponds to $p-q$ (see section 5.2.2). In the following, we replace the translation group $\boldsymbol{T}$ by this additive group. If the neighborhood structure is infinite then the notions introduced are based on the properties of translative neighborhood structures (see section 4.1.3).

But if we take $[P,N]$ as finite then the neighborhood structure is toroidally closed. Then the translation group is finite, and we have a finite group order because $P = P_1 \times P_2 \times \ldots \times P_k$ is the factorization of P as a direct product of cyclical subgroups. Thereby, the number k of the cyclical subgroups $P_1, \ldots, P_k$ are connected in any way with the dimension n of the neighborhood structure and with the periodicities of $[P,N]$.

Discrete functions can be neither differentiated nor integrated. However, the functions (i.e. their values) can be added, subtracted, and multiplied by addition, subtraction, and multiplication of their coefficients. In general, divisions do not occur. The finiteness of the domain of definition together with the exclusion of general divisions suggests rings, fields and algebras as suitable algebraic structures to describe discrete functions.

The discrete Fourier transform (DFT) and the FFT algorithm were investigated in finite fields and rings already in the past [Ni71, Po71]. The FFT algorithm applications to Fourier, Walsh, and Hadamard transforms were shown to be a special case of Fourier transform in algebra over fields or rings. Over the last twenty years, attention has been focused on NNT's over modulo rings $N(m)$ where m is a positive integer [Ag75, Nu81, Ra72, Re75].

The common goal of this research is characterized by the development of fast algorithms and transforms and the construction of corresponding computer architecture. Here we want to generalize the algebraic approach for computing circular convolutions based on the above mentioned investigations without regard to machine realization [Vo89a, Vo91c].

Our aim is to show the deep agreement and conformity of all these methods as demonstrable by algebraic methods.

Def. 5.2.-1: The set $\boldsymbol{F}$ of all mappings F from P of $[P,N]$ into a commutative ring R with unit element e and without zero divisors (i.e. R is a *domain of integrity* - see section 5.2.2) is called *discrete function space*, and its elements are called *discrete functions*.

For instance, the sets of integer numbers, real numbers, or complex numbers are domains of integrity. We use the following properties of the both operations *addition* of two functions and *multiplication* of a function by a ring element where F_p means the function value for $p \in P$:

$$H_p = (F+G)_p = F_p + G_p \;, \qquad F,G,H \in \boldsymbol{F} \;, \qquad p \in P$$

$$H_p = (aF)_p = aF_p \;, \qquad F,H \in \boldsymbol{F} \;, \quad p \in P \;, \quad a \in R \;.$$

With these assumptions, we can treat the function space $\boldsymbol{F}$ as a finite dimensional vector space. The discrete elementary functions $E^{(q)} \in \boldsymbol{F}$ defined by

$$E_p^{(q)} = \begin{cases} e & \text{for} \quad p = q \\ o & \text{for} \quad p \neq q \end{cases} \qquad \text{with} \quad p,q \in P$$

with e as unit element of the domain of integrity and o as zero element represent a *function base* in $\boldsymbol{F}$. Indeed, it can be derived now that the equation

$$F = \sum_{q \in P} a_q E^{(q)} \qquad , \qquad a_q \in R$$

has an unambiguous solution if we introduce the *pseudo scalar product* (shortly *P-product*) which fulfills the distributive law $<F+G,H>=<F,H>+<G,H>$:

$$\langle F , G\rangle = \sum_{p\in P} F_p G_p \ , \qquad F, G \in \boldsymbol{F} \ .$$

Then, it follows

$$\langle E^{(p)} , F\rangle = \langle E^{(p)} , \sum_{q\in P} a_q E^{(q)} \rangle$$

$$= \sum_{q\in P} a_q \langle E^{(p)} , E^{(q)} \rangle = \sum_{q\in P} a_q \delta_{p-q} = a_p$$

where δ_{p-q} is the *Kronecker symbol* (equal to the unit element e for $p=q$ and equal to the zero element o for $p\neq q$). We call the set of all functions $E^{(p)}$ a *canonical base* of $\boldsymbol{F}$.

5.2.2 Algebraic Structures

We will now shortly discuss the main types of algebraic structures used in image processing. Like set theory, graph theory, or theory of relations, the theory of algebraic structures is a large mathematical discipline. Only the basic notions shall be compiled here because there are a lot of related textbooks. Therefore, we will omit the proofs for some of the theorems quoted here.

Def. 5.2.-2: Be given a set A of elements. A mapping from the set product $A^k = A\times A\times \ldots \times A$ to A is called a k-ary algebraic combination C, and we write $C(a_1,\ldots,a_k)=a_r$. The element $a_r\in A$ is the *result* of the algebraic combination C of elements $a_1,\ldots,a_k$. For $k=2$, we speak of a *binary* algebraic combination (shortly *algebraic combination*). A binary combination $C(a,b)=c$ of two elements $a,b\in A$ is symbolized by $a\circ b=c$.

Of course, we can consider a k-ary algebraic combination as a $(k+1)$-ary relation $(a_1,\ldots,a_k,a_r)\in A^{k+1}$ (see also section 1.1.2). But contrary to general relations, all of that $(k+1)$-tupels $(a_1,\ldots,a_k,a_r)$ are elements of the relation for which $(a_1,\ldots,a_k)\in A^k$. Corresponding to binary algebraic combinations, the result $c=a\circ b$ is defined for any elements $a,b\in A$.

Therefore, we introduce new notions for the theory of algebraic combinations although these notions have connections to the notions of the theory of relations. For instance, if from $(a,b)\in R\subseteq A^2$ always $(b,a)\in R$ follows, we say that R is a symmetrical relation. Otherwise, we say that a binary algebraic combination is *commutative* if $C(a,b,c)=C(b,a,c)$ or $a\circ b=b\circ a$.

A binary algebraic combination is *associative* if $(a \circ b) \circ c = a \circ (b \circ c)$. We will use only such algebraic combinations in our further investigations and discussions which are binary, commutative, and associative. Because of that, we will speak simply of algebraic combinations.

Def. 5.2.-3: An *algebraic structure* is a non-empty set A of elements with one or some algebraic combinations defined for the elements of A.

Def. 5.2.-4: Any set G together with a binary combination is a *groupoid*. A groupoid whose binary combination associative is named a *semigroup*. An algebraic structure $[A, e, \circ]$ is denoted as *monoid* if the combination $\circ$ fulfills the associative law and if there is an unit element e.

Def. 5.2.-5: A (commutative) *group* is a set G of elements with an algebraic combination for which the following axioms are fulfilled:

- It is $a \circ b = b \circ a$ and $(a \circ b) \circ c = a \circ (b \circ c)$.
- There is an element e of G (called *unit element*) so that $e \circ a = a \circ e = a$ for each $a \in G$.
- For each $a \in G$ there exists an *inverse* element $a^{-1} \in G$ so that $a \circ a^{-1} = a^{-1} \circ a = e$.

As examples for groups, we can use the sets of integer or real numbers with addition as combination. The relationships $(x+y)+z=x+(y+z)$ with $+$ instead of $\circ$ as combination sign, $0+x=x+0=x$ with 0 as "unit element", and $x+(-x)=0$ with $-x$ as "inverse element" of x are fulfilled.

Groups can consist from a finite number of elements. The set $G=\{\boldsymbol{I},\boldsymbol{C}\}$ of the two elements $\boldsymbol{I}$ (identity operator in neighborhood structures, see section 1.4.3) and $\boldsymbol{C}$ (complement operator) is a finite group with successive application of operators as algebraic combination. The identity operator is the unit element of the group, and the inverse element of $\boldsymbol{C}$ is $\boldsymbol{C}^{-1}=\boldsymbol{C}$.

The number of elements in a finite group G is the *order* O(G) of the group. For groups of a small order, it is possible to write down all combinations in form of a *group table*.

	e	a
e	e	a
a	a	e

	e	a	b
e	e	a	b
a	a	b	a
b	b	e	a

	e	a	b
e	e	a	b
a	a	e	e
b	b	e	e

We see that the third of these tables does not describe a group because $(a \circ b) \circ b = b$ is different from $a \circ (b \circ b) = a$ so that the associative law is violated. It is a task typically for algebraic problems to ask for the number of different groups of order k. In the case of $k=4$, there are only two different groups:

	e	a	b	c
e	e	a	b	c
a	a	b	c	e
b	b	c	e	a
c	c	e	a	b

	e	a	b	c
e	e	a	b	c
a	a	e	c	b
b	b	c	e	a
c	c	b	a	e

Def. 5.2.-6: If a subset $G' \subset G$ fulfills all group axioms, we call G' a *subgroup* of G. The powers of any element $a \in G$ together with the powers of a^{-1} and with the unit element e are always a subgroup G_a which we denote as the *cyclic subgroup* generated by a.

Theorem 5.2.-1: The order of a subgroup of a finite group is always a divisor of the order of the group.

Now, we will introduce algebraic structures with two different combinations, namely rings and fields. For rings, we have the set of integers as usual example, and for fields, the set of real numbers can be used to illustrate the axioms and theorems. But it is to be taken into account that there are also finite rings and finite fields.

Def. 5.2.-7: A (commutative) *ring* is a set R of elements with two algebraical combinations denoted by $\oplus$ and $\odot$ for which the following axioms are fulfilled:

- In respect of combination $\oplus$ denoted as *addition*, the elements of R are an (additive) group whereby we call here the unit element of the group as *null element* o.
- In respect of combination $\odot$ denoted as *multiplication*, the elements of R fulfill the associative law $a \odot (b \odot c) = (a \odot b) \odot c$.
- The relationships between the two combinations $\oplus$ and $\odot$ are given by the distributive law $a \odot (b \oplus c) = (a \odot b) \oplus (a \odot c)$.

A ring contains always a null element. But it is not necessary that a ring contains a unit element e with $a \odot e = a$ (for instance, the ring of all even numbers does not contain a unit element). Further, it is possible in a ring that $a \odot b = o$ although neither a nor b are identical with the null element o. We denote an element $a \in R$ as *zero divisor* if there are elements $b \in R$ so that $a \odot b = o$.

The ring of all 2·2-matrices with integer coefficients contains zero divisors. If we use the common addition and multiplication rules for matrices as algebraic combinations, the two elements

$$e = \begin{pmatrix} 1 & 0 \\ 0 & 1 \end{pmatrix} \quad , \qquad o = \begin{pmatrix} 0 & 0 \\ 0 & 0 \end{pmatrix}$$

are the unit element *e* and the null element *o* of this ring. It is easy to see that this ring of matrices contains zero divisors because for instance

$$\begin{pmatrix} 1 & -1 \\ -1 & 1 \end{pmatrix} \begin{pmatrix} 1 & 1 \\ 1 & 1 \end{pmatrix} = \begin{pmatrix} 0 & 0 \\ 0 & 0 \end{pmatrix} .$$

A commutative ring with unit element *e* but without zero divisors is a *domain of integrity*.

An important example of a domain of integrity is the ring of *polynomials* $P(x)$ of unknown x, i.e. any two polynomials

$$P(x) = \sum_{k=0}^{\infty} p_k x^k \quad , \qquad Q(x) = \sum_{k=0}^{\infty} q_k x^k$$

can be added, subtracted and multiplicated with one another. The coefficients of the product $R(x)=P(x)\cdot Q(x)$ are given by

$$r_k = \sum_{l=0}^{k} p_l \, q_{k-l} \ .$$

This formula emphasizes the interest which is spend to polynomial rings in image processing because the convolution of functions is defined in a very similar way (see section 5.1.3).

Def. 5.2.-8: A (commutative) *field* is a set S of elements with two algebraic combinations denoted by $\oplus$ and $\odot$ for which the following axioms are fulfilled:

- In respect of combination $\oplus$ denoted as *addition*, the elements of S are an (additive) group whereby we call here the unit element of the group the *null element o*.
- In respect of combination $\odot$ denoted as *multiplication*, the elements of S fulfill both, the commutative law $a \odot b = b \odot a$ and the associative law $a \odot (b \odot c) = (a \odot b) \odot c$, and for all elements $a \neq o$ exist an inverse element a^{-1} with $a \odot a^{-1} = e$, i.e. the set $S - \{o\}$ is a *multiplicative group*.
- The relationships between the two combinations $\oplus$ and $\odot$ are given by the distributive law $a \odot (b \oplus c) = (a \odot b) \oplus (a \odot c)$.

In fields, we have always a null element o with respect to the addition $\oplus$ and a unit element e in respect of the multiplication $\odot$. Further, the product $a\odot b$ of two elements is then and only then equal to the null element o if at least one of the two elements is the null element. The proof is simple: Either is $a=o$ in $a\odot b=o$ or a is different from o. In this case a^{-1} exists in the field so that

$$b = e\odot b = \left(a^{-1}\odot a\right)\odot b = a^{-1}\odot o = o$$

follows.

The set of real numbers and the set of complex numbers are examples for fields. Fields with a finite number of elements are called *Galois fields*. The most simple example for a Galois field is the field which consists only of the two elements e and o. The two field tables are given by

$\oplus$	o	e
o	o	e
e	e	o

and

$\odot$	o	e
o	o	o
e	o	e

.

If we interpret the two elements as Boolean variables o (false) and e (right) and use for the "addition" $\oplus$ the logical XOR (exclusive or) as operation, for the "multiplication" $\odot$ the logical AND, then we obtain a realization of a Galois field with only two elements.

Theorem 5.2.-2: Each finite domain of integrity is a field.

The proof of this theorem takes into account, that in a finite domain of integrity all elements can be numbered as $a_0=o,a_1=e,\ldots,a_{N-2}$ with N as number of elements. For any element $a\neq o$, the products $a\odot a_k$ and $a\odot a_l$ must be different if $k\neq l$ because we have assumed that $a\odot(a_k-a_l)\neq o$ for $a_k-a_l\neq o$. Therefore, the N different products $a\odot a_k$ for $a\neq o$ yield N different results, and exactly one of these results must be the unit element e so that from $a\odot a_m=e$ the existence of $a^{-1}=a_m$ follows.

Def. 5.2.-9: Be $A=\{a,b,\ldots\}$ an (associative) ring with the two combinations $\oplus$ and $\odot$. Further let $R=\{\alpha,\beta,\ldots\}$ be a ring with unit element and with combinations $\cdot$ and $+$. We call A an *algebra over R* or a *hypercomplex system over R* if there is a further multiplicative combination $\alpha\circledast a\in A$ and if the following conditions are fulfilled:

- $\alpha\circledast(a\oplus b)=(\alpha\circledast a)\oplus(\alpha\circledast b)$, $(\alpha+\beta)\circledast a=(\alpha\circledast a)\oplus(\beta\circledast a)$,
- $\alpha\circledast(a\odot b)=(\alpha\circledast a)\odot b$, $(\alpha\cdot\beta)\circledast a=\alpha\circledast(\beta\circledast a)$.

For instance, A can be the ring of 2·2-matrices with integer coefficients with matrix addition $\oplus$ and matrix multiplication $\odot$, and R can be the ring of integers with common addition + and common multiplication $\cdot$. The multiplication $\circledast$ of a matrix $a=(a_{ik})\in A$ with an integer $\alpha\in R$ is then defined as coefficient-wise multiplication, i.e. $\alpha\circledast a=\alpha\circledast(a_{ik})=(\alpha\cdot a_{ik})$.

5.2.3 Convolution of Functions

Now, we will investigate discrete functions $F\in\boldsymbol{F}$ with coefficients F_p with respect to the foregoing algebraic notions and definitions. The coefficients shall be elements of a ring R, and the domain of definition P shall be a translative neighborhood structure.

To begin, we introduce a new operation

$$F*G = H , \qquad H_p = \sum_{q\in P} F_{p-q}\cdot G_q , \qquad F,G,H\in \boldsymbol{F}$$

which we denote as (circular) *convolution* of the two functions F and G. The convolution is an operation in $\boldsymbol{F}$ in that sense that $F*G$ is always an element of $\boldsymbol{F}$. This operation has the following properties:

$$F*G = G*F \qquad \text{commutativity ,}$$

$$(F*G)*H = F*(G*H) \qquad \text{associativity ,}$$

$$F*(G+H) = F*G + F*H \qquad \text{distributivity ,}$$

$$a\,(F*G) = (a\,F)*G \qquad \text{compatibility .}$$

Here, a is any element of a ring R' whereby R' can be the same ring from which the coefficients of the functions are taken.

Together with the coefficient-wise additivity of functions $F,G\in\boldsymbol{F}$, it follows from these properties that $\boldsymbol{F}$ is a *commutative algebra* over the ring R where R is the ring of function values. For two functions

$$F = \sum_{p\in P} \varphi_p B^{(p)} , \qquad G = \sum_{q\in P} \gamma_q B^{(q)} , \qquad \varphi_p , \gamma_q \in R$$

represented as sums over any function base $\{B^{(p)}, p\in P\}$ especially the canonical base of elementary unit functions $E^{(p)}$, we obtain the general expression

$$F*G = \sum_{p\in P} \varphi_p B^{(p)} * \sum_{q\in P} \gamma_q B^{(q)} = \sum_{p\in P} \sum_{q\in P} \varphi_p \gamma_q (B^{(p)} * B^{(q)}) .$$

Otherwise, the convolutions $B^{(p)}*B^{(q)}$ are also elements of $\boldsymbol{F}$ and can be expressed as linear combinations of functions $B^{(p)}$:

$$B^{(p)} * B^{(q)} = \sum_{r\in P} c_{pq}^{r} B^{(r)} \quad , \qquad c_{pq}^{r} \in R \ .$$

Therefore, we obtain

$$F*G = \sum_{p\in P} \sum_{q\in P} \sum_{r\in P} \varphi_p \gamma_q c_{pq}^{r} B^{(r)} \ .$$

The c's are called *structure constants* of the algebra $\boldsymbol{F}$ in respect of the base $\{B^{(p)}, p\in P\}$. For any given base, the structure constants are unambiguously determined. Especially for the canonical base $\{E^{(p)}, p\in P\}$, it follows

$$c_{pq}^{r} = \left\langle E^{(p)} * E^{(q)} \ , \ E^{(r)} \right\rangle = \begin{cases} e & \text{for} \quad p+q=r \\ o & \text{for} \quad p+q\neq r \end{cases}$$

where the brackets denote the P-product introduced in section 5.2.-1. The relationship between P-product and convolution is given by

<u>Formula 5.2.-1:</u>

$$\begin{aligned} \langle F*G, H\rangle &= \sum_{p\in P} (F*G)_p H_p = \sum_{p\in P} \sum_{q\in P} F_{p-q} G_q H_p \\ &= \sum_{q\in P} G_q \sum_{p\in P} F_{p-q} H_p = \sum_{q\in P} G_q \sum_{p\in P} F^{\leftrightarrow}_{q-p} H_p \\ &= \sum_{q\in P} G_q \, (F^{\leftrightarrow} * H)_q = \langle G, F^{\leftrightarrow} * H\rangle \ . \end{aligned}$$

Here, $F^{\leftrightarrow}$ is the *reflected function* corresponding to F which can be generally defined (see section 5.1.2) because the element $-p\in P$ always exists in a translative neighborhood structure $[P,N]$.

5.2.4 Convolution Orthogonality

The theory of commutative algebras states that a commutative ring R and the structure constants $c_{p\,q}^{\,r}$ determine unambiguously a commutative algebra. The group properties of translations in the neighborhood structure $[P,N]$ establish the structure constants.

In a given space $\boldsymbol{F}$, a transform of base elements $B^{(p)}$ to other base elements $C^{(q)}$ induces likewise a change of structure constants. Up to now, the structure constants are known for the canonical base $\{E^{(p)}, p\in P\}$. Otherwise, the number of general additions and multiplications needed to calculate the convolution of two functions is proportional to the number of nonzero structure constants. In the case of the canonical base, we have $(card(P))^2$ nonzero c's. However, the following structure constants are yet more simple:

$$c_{pq}^{r} = \begin{cases} e & \text{for} \quad p=q=r \\ o & \text{else} \end{cases} .$$

Such structure constants require only *card*(*P*) additions and multiplications for the numerical evaluation of convolution products. The function base corresponding to these simple c's, provided it exists, is called an *orthogonal convolution base* (OCB). In OCB's, the basic functions $C^{(p)}$ have the property of *convolution orthogonality*

$$C^{(p)} * C^{(q)} = \sum_{p,q,r \in P} c_{pq}^{r} C^{(r)} = \delta_{p-q} C^{(p)} .$$

A further important fact is that from the convolution orthogonality also follows a "common" orthogonality described by the *P*-product:

$$\begin{aligned}(C^{(p)} * C^{(q)})_0 &= \delta_{p-q} (C^{(p)})_0 \\ &= \sum_{r \in P} (C^{(p)})_{-r} (C^{(q)})_r = \langle (C^{(p)})^{\leftrightarrow}, C^{(q)} \rangle .\end{aligned}$$

This equation expresses that the orthogonality between two functions in respect of the *P*-product is a conclusion of the convolution orthogonality.

Finally, the general convolution of two functions is given by the very simple expression

<u>Formula 5.2.-2:</u>

$$F*G = \sum_{p \in P} \varphi_p C^{(p)} * \sum_{q \in P} \gamma_q C^{(q)} = \sum_{p \in P} \varphi_p \gamma_p C^{(p)} .$$

This equation is equivalent to the generalization of the known *discrete convolution theorem* for Fourier transforms according to which the Fourier coefficients of the convolution of two functions F and G are the products of Fourier coefficients φ_p and γ_p of the single functions. Using the discrete convolution theorem, a convolution of two functions is mapped into pointwise products of their coefficients.

The existence of an orthogonal convolution base guarantees the validity of the discrete convolution theorem (DCT). Vice versa, the validity of DCT demands the existence of an orthogonal convolution base (OCB). Therefore, it is necessary to search for OCB's in function spaces $\boldsymbol{F}$ to simplify numerical convolutions needed in digital image processing.

There exists an effective solution to this problem based on the properties of rings. Let ***F*** be a commutative algebra over a domain of integrity *R* with identity *e*. Furthermore, let *P* be the commutative *cyclic group* of order *N* corresponding to a (one-dimensional) neighborhood structure with *N* points. Then, the following theorem is valid:

Theorem 5.2.-3: An OCB exists in ***F*** if and only if

$$a^N = e \;,$$

$$(a^k - e)^{-1} \quad \text{exists for} \quad k \in \{1, \ldots, N-1\} \;,$$

$$N^{-1} = (Ne)^{-1} = (e+e+\ldots+e)^{-1} \quad \text{exists} \;,$$

i.e. the algebraic quantity $a \epsilon R$ is a primitive *N*-th root of the unit element *e* of *R*, and the expressions $a^k - e$ are invertible in *R*.

These conditions are sufficient for the existence of OCB's, and therefore they are also sufficient for the existence of DCT's in general Fourier transforms as it was proven in many different papers over the past years. But these conditions are also necessary.

In the special case of the one-dimensional cyclic lattice, the conditions of the theorem regarding the existence of OCB are very simple because the *p*'s are mapped to integers $0 \ldots N-1$. The one-dimensional functions $C^{(n)}$ of the OCB are representable by their coefficients

$$C_l^{(k)} = N^{-1} a^{kl} \quad \text{for} \quad k, l \in \{0, \ldots, N-1\} \;.$$

It will be the task of the following sections to give examples of domains of integrity and examples of OCB's.

5.3 Orthogonal Convolution Bases

5.3.1 General Properties in OCB's

We will investigate in this chapter the general properties of functions of orthogonal convolution bases (OCB's). In connection with the development of discrete signal theory - i.e. the amplitudes of signals are integer values - *generalized Fourier transforms* over finite fields and commutative rings were introduced in the past [Ni71, Du78, Du80].

The algebraic quantity a introduced in theorem 5.2.-3 and the period length N of a one-dimensional periodical discrete function are the main characteristics of the theory of OCB's. For clarity, now we will denote a primitive N-th root of the unit element e by $\boldsymbol{a}$.

A formula essential for many further investigations is given by

$$x_m = \sum_{k=0}^{N-1} \boldsymbol{a}^{km} = N\delta_m$$

with $\delta_m = e$ for $m=0$ and $\delta_m = o$ for $m \neq 0$ (e is the unit element and o is the zero element of the corresponding algebraic structure). In fact, the formula is valid for $m=0$. Further, we obtain

$$(\boldsymbol{a}^m - e)\, x_m = \sum_{k=0}^{N-1} (\boldsymbol{a}^{k(m+1)} - \boldsymbol{a}^{km}) = \sum_{k=1}^{N} \boldsymbol{a}^{km} - \sum_{k=0}^{N-1} \boldsymbol{a}^{km}$$

$$= \boldsymbol{a}^{Nm} - \boldsymbol{a}^0 = (\boldsymbol{a}^N)^m - e = o \,.$$

If we use the assumption that $(\boldsymbol{a}^m - e)^{-1}$ exists for $m \neq 0$, we see that $x_m = o$ for $m \neq 0$. Therefore we obtain the following theorem:

<u>Theorem 5.3.-1:</u> The general structure of basic functions of an orthogonal convolution base is

$$C^{(k)} = N^{-1}(e, \boldsymbol{a}^k, \boldsymbol{a}^{2k}, \ldots, \boldsymbol{a}^{(N-1)k}) \,.$$

The coefficients of these functions satisfy the formulas $C_n^{(k)} = N^{-1} \cdot \boldsymbol{a}^{kn}$ so that the convolution between two functions $C^{(k)}$ and $C^{(l)}$ is

$$(C^{(k)} * C^{(l)})_n = \sum_{m=0}^{N-1} C_m^{(k)} C_{n-m}^{(l)} = N^{-2} \sum_{m=0}^{N-1} \boldsymbol{a}^{km} \boldsymbol{a}^{ln-lm}$$

$$= N^{-2} \boldsymbol{a}^{ln} \sum_{m=0}^{N-1} \boldsymbol{a}^{m(k-l)} = N^{-2} \boldsymbol{a}^{ln} N \delta_{k-l} = \delta_{k-l} C_n^{(l)} \,.$$

Therefore, it follows

$$C^{(k)} * C^{(l)} = \delta_{k-l} C^{(l)} \quad .$$

We can represent each function $F \epsilon \boldsymbol{F}$ as sum

$$F = \sum_{k=0}^{N-1} \varphi_k C^{(k)}$$

over the base functions $C^{(k)}$. The coefficients φ_k can be determined by multiplication of F_n with $\boldsymbol{a}^{-nl}$ and summation over all n:

$$\sum_{n=0}^{N-1} F_n \boldsymbol{a}^{-nl} = N^{-1} \sum_{n=0}^{N-1} \sum_{k=0}^{N-1} \varphi_k \boldsymbol{a}^{n(k-l)} = \varphi_l \quad .$$

In this way, a pair of formulas arises for the transform and *re-transform* between function coefficients F_n and *transform coefficients* φ_k:

<u>Formula 5.3.-1:</u>

$$F_n = N^{-1} \sum_{k=0}^{N-1} \varphi_k \boldsymbol{a}^{kn} \quad ,$$

$$\varphi_k = \sum_{n=0}^{N-1} F_n \boldsymbol{a}^{-kn} = N \langle F, (C^{(k)})^{\leftrightarrow} \rangle \quad .$$

<u>Theorem 5.3.-2:</u> The functions $C^{(k)}$ of an OCB are eigen functions to each convolution operator $\boldsymbol{C_F}$ corresponding to function F, and the eigen values are given by φ_k.

The proof of this theorem is relatively simple. It is

$$(\boldsymbol{C_F} C^{(k)})_l = (F * C^{(k)})_l = \sum_{m=0}^{N-1} F_m \cdot C_{l-m}^{(k)} = N^{-1} \sum_{m=0}^{N-1} F_m \cdot \boldsymbol{a}^{(l-m)k}$$

$$= N^{-1} \boldsymbol{a}^{kl} \sum_{m=0}^{N-1} F_m \cdot \boldsymbol{a}^{-mk} = N^{-1} \varphi_k \boldsymbol{a}^{kl} = \varphi_k C_l^{(k)} \quad .$$

This result means that $(F*C^{(k)})_l = \varphi_k C_l^{(k)}$ or $F*C^{(k)} = \varphi_k C^{(k)}$, i.e. the base functions $C^{(k)}$ are eigen functions to each convolution operator $\boldsymbol{C_F}$ whereby only the eigen values are different.

If the transform coefficients φ_k and γ_k of two functions F and G are known, we obtain the convolution of the functions as

$$H = F * G = \sum_{k=0}^{N-1} \sum_{l=0}^{N-1} \varphi_k \gamma_l C^{(k)} * C^{(l)} = \sum_{k=0}^{N-1} \varphi_k \gamma_k C^{(k)} .$$

This is the so-called *convolution theorem* that the transform coefficients χ_k of function $H=F*G$ are given by the simple product

$$\chi_k = \varphi_k \gamma_k \quad \text{for} \quad k \in \{0,1,\ldots,N-1\} .$$

We will symbolize this *generalized Fourier transform* by the sign $\boldsymbol{F}$, i.e. $\boldsymbol{F}(F)$ means the tupel $(\varphi_0,\varphi_1,\ldots,\varphi_{N-1})$ instead of the tupel $F=(f_0,f_1,\ldots,f_{N-1})$. Some general properties of the generalized Fourier transform are the following:

- *Linearity*, i.e. $\boldsymbol{F}(F+G) = \boldsymbol{F}(F)+\boldsymbol{F}(G)$,

- *Convolution property*, i.e. $\boldsymbol{F}(F*G) = \boldsymbol{F}(F)\cdot\boldsymbol{F}(G)$ whereby the product means the coefficient-wise multiplication of $\boldsymbol{F}(F)$ and $\boldsymbol{F}(G)$,

- *Shift property*, i.e. $\boldsymbol{F}(S^{|k}F) = N\cdot C^{(-k)}\cdot\boldsymbol{F}(F)$,

- *Reflection property*, i.e. $\boldsymbol{F}(\boldsymbol{R}F) = \boldsymbol{R}\boldsymbol{F}(F)$ whereby the reflection of a function X means $(\boldsymbol{R}X)_l=X_{-l}$.

5.3.2 Fourier Transform

There is the question whether an algebraic structure exists so that the periodicity condition $\boldsymbol{a}^N=e$, the existence of $(\boldsymbol{a}^k-e)^{-1}$, and the existence of N^{-1} are fulfilled.

In common arithmetic, it is not possible to give a positive answer to this question. For instance in the case of $N=3$, we have to search for such a function $X=(x_0,x_1,x_2)$ that the convolution equation $X*X=X$ is fulfilled. We obtain

$$\begin{aligned}(x_0,x_1,x_2)*(x_0,x_1,x_2) &= (x_0^2+2x_1x_2,\, x_2^2+2x_0x_1,\, x_1^2+2x_2x_0) \\ &= (x_0,x_1,x_2) .\end{aligned}$$

The convolution equation is equivalent to the nonlinear equation system

$$\begin{aligned} x_0^2 + 2x_1x_2 &= x_0 , \\ x_2^2 + 2x_0x_1 &= x_1 , \\ x_1^2 + 2x_2x_0 &= x_2 . \end{aligned}$$

This system is solvable with real numbers, as the solutions (2/3,-1/3,-1/3) or (1/3,1/3,1/3) for X show. But we have also to search for three different functions X_0,X_1,X_2 so that always $X_k{*}X_l=\delta_{k-l}\,X_l$, and these conditions yield

$$X_0 = \frac{1}{3}(1,1,1),$$

$$X_1 = \frac{1}{3}(1, e^{2\pi i/3}, e^{4\pi i/3}),$$

$$X_2 = \frac{1}{3}(1, e^{4\pi i/3}, e^{2\pi i/3}).$$

Now, it is possible to fulfill all algebraic requirements in the field of complex numbers with $e=1$. We take

$$\boldsymbol{a} = e^{\frac{2\pi i}{N}} \qquad \text{with} \quad \boldsymbol{a}^N = 1 .$$

Then

$$(\boldsymbol{a}^k - 1)^{-1} = \frac{1}{e^{\frac{2\pi i k}{N}} - 1} = \frac{\left(\cos\frac{2\pi k}{N} - 1\right) - i\sin\frac{2\pi k}{N}}{2\left(1 - \cos\frac{2\pi k}{N}\right)}$$

exists for $k \neq 0$ *modulo* N, and finally N^{-1} exists for all finite positive values N.

The functions

$$C^{(k)} = N^{-1}\left(1, e^{\frac{2\pi i\cdot k}{N}}, e^{\frac{2\pi i\cdot 2k}{N}}, \ldots, e^{\frac{2\pi i\cdot (N-1)k}{N}}\right)$$

constitute the orthogonal convolution base whereby $C^{(k)}{*}C^{(l)}=\delta_{k-l}\,C^{(l)}$. Finally, we have

Formula 5.3.-2:

$$F_n = \frac{1}{N}\sum_{k=0}^{N-1} \varphi_k\, e^{\frac{2\pi i\cdot kn}{N}} , \qquad \varphi_k = \sum_{n=0}^{N-1} F_n\, e^{-\frac{2\pi i\cdot kn}{N}}$$

for the Fourier transform and the inverse Fourier transform, respectively.

All properties of the general transform formulated for the algebraic quantity $\boldsymbol{a}$ are fulfilled also for this special case of complex numbers. But there are of course further properties caused by the laws of complex numbers. For instance, the reflection of a real valued function is given by $\boldsymbol{F}(\boldsymbol{R}F)=(\boldsymbol{F}(F))^*$ where $(\ldots)^*$ means complex conjugated numbers.

Because $F(F)=\varphi=(\varphi_0,\varphi_1,\ldots,\varphi_{N-1})$ is also a function, we can transform it to $\psi=F\varphi$ with

$$\psi_l = \sum_{k=0}^{N-1} \varphi_k \, a^{-kl} = \sum_{k=0}^{N-1}\sum_{n=0}^{N-1} F_n \, a^{-k(n+l)} = N \cdot F_{-l} \ .$$

Therefore, the two-fold Fourier transform $F(F(F))$ yields the reflected function RF up to the factor N.

Two-dimensional Fourier transforms can be described by

$$F_{nm} = \frac{1}{N}\frac{1}{M}\sum_{k=0}^{N-1}\sum_{l=0}^{M-1} \varphi_{kl} \, a_1^{nk} \, a_2^{ml}$$

with two algebraic quantities a_1 and a_2 which fulfills the conditions

$$a_1^N = e = 1 \quad , \qquad a_2^M = e = 1 \ ,$$

and the other requirements. Because of

$$F_{nm} = \frac{1}{N}\sum_{k=0}^{N-1}\left(\frac{1}{M}\sum_{l=0}^{M-1}\varphi_{kl}\, e^{\frac{2\pi i\cdot ml}{M}}\right) e^{\frac{2\pi i\cdot nk}{N}} = \frac{1}{N}\sum_{k=0}^{N-1}\psi_{km}\, e^{\frac{2\pi i\cdot nk}{N}}$$

the separation of the two-dimensional transform in two single one-dimensional transforms is possible.

It is not the task here to develop the full theory of Fourier transform. The aim was only to show that the Fourier transform is a special case of the general algebraic transform based on orthogonal convolution bases.

5.3.3 Number Theoretical Transforms

We will return to the main topic of this chapter, namely to the discrete functions. Although the Fourier transform yields a solution of the algebraic problem $a^N=e$, it uses complex numbers in contradiction to the fact that in image processing all function values should be integers if possible.

We can work with other algebraic quantities such as $\exp(2\pi i/N)$ because we have only to guarantee that the conditions $a^N=e$, existence of $(a^k-e)^{-1}$, and existence of N^{-1} are fulfilled. All other problems connected with representations and transforms of functions are caused by practical reasons and numerical requirements.

It is not sufficient to take into account the first two requirements. The relationship $a^N=e$ and the existence of $(a^k-e)^{-1}$ are fulfilled for cyclic subgroups of order N with $\boldsymbol{a}$ as generating element. For instance, the permutations

$$\boldsymbol{a} = \begin{pmatrix} a & b & c & d & e & f & g & h & i & j \\ b & a & e & c & d & j & f & g & h & i \end{pmatrix}$$

$$e = \begin{pmatrix} a & b & c & d & e & f & g & h & i & j \\ a & b & c & d & e & f & g & h & i & j \end{pmatrix}$$

yield the primitive period N=30=2·3·5 as the following examples show:

$$\boldsymbol{a}^2 = \begin{pmatrix} a & b & c & d & e & f & g & h & i & j \\ a & b & d & e & c & i & j & f & g & h \end{pmatrix},$$

$$\boldsymbol{a}^3 = \begin{pmatrix} a & b & c & d & e & f & g & h & i & j \\ b & a & c & d & e & h & i & j & f & g \end{pmatrix} \quad \text{etc.}$$

But the problem is that there are no laws as to how we have to add $\boldsymbol{a}^2$ and $\boldsymbol{a}^3$ or as to how we have to multiply a permutation with an integer number. Permutations are elements of a group, not elements of a ring.

An algebraic structure better suited for our purposes is given by matrix algebra. The matrix

$$\boldsymbol{a} = \begin{pmatrix} 0 & -1 & 0 & 0 & 0 & 0 & 0 & 0 \\ 0 & 0 & 1 & 0 & 0 & 0 & 0 & 0 \\ 1 & 0 & 0 & 0 & 0 & 0 & 0 & 0 \\ 0 & 0 & 0 & 0 & 1 & 0 & 0 & 0 \\ 0 & 0 & 0 & 0 & 0 & 1 & 0 & 0 \\ 0 & 0 & 0 & 0 & 0 & 0 & 1 & 0 \\ 0 & 0 & 0 & 0 & 0 & 0 & 0 & 1 \\ 0 & 0 & 0 & 1 & 0 & 0 & 0 & 0 \end{pmatrix}$$

fulfills the condition $\boldsymbol{a}^N = e$ with N=30=6·5 where e is the unit matrix. We are equipped here with a law of addition $\boldsymbol{a}^2+\boldsymbol{a}^3$ of matrices and with a law of multiplication of a matrix with integer or real numbers.

The disadvantages of matrix algebra with respect to numerical realization are the large memory requirements and the cumbersome treatment of matrix operations in comparison to simple number operations.

These disadvantages can be avoided by means of number theoretical transforms (NTT, see [Ag75, Mc79, Th83]). As algebraic quantities $\boldsymbol{a}$, we use residue classes c in respect of a given modul M, i.e. to a given integer number. We will show here an example with $M=11$ and $\boldsymbol{a}=3$. It is

$$\begin{aligned} \boldsymbol{a}^1 &\equiv 3 \quad \mathit{modulo}\ 11\ , \\ \boldsymbol{a}^2 &\equiv 9 \quad \mathit{modulo}\ 11\ , \\ \boldsymbol{a}^3 &\equiv 5 \quad \mathit{modulo}\ 11\ , \\ \boldsymbol{a}^4 &\equiv 4 \quad \mathit{modulo}\ 11\ , \\ \boldsymbol{a}^5 &\equiv 1 \quad \mathit{modulo}\ 11\ . \end{aligned}$$

Therefore the periodicity condition $\boldsymbol{a}^5=e=1$ is fulfilled. The number theoretic inverse to $N=5$ is $N^{-1}=9$ because $5\cdot 9\equiv 1$ *modulo* 11 (we will use here the sign $\equiv$ for the equivalence of two residue classes).

The advantages of number theoretic algebraic structures are mainly that there are very simple laws of addition and multiplication like the laws in integer arithmetic. But contrary to integer arithmetic, rings of residue classes are finite algebraic structures and there are only residue classes and integer numbers so that they yield excellent tools for treatment of discrete periodic functions with integer coefficients.

In general, it is not necessary that the algebraic structure has to be a finite field, i.e. a *Galois field*. We need only the assumption that the requirements of theorem 5.2.-3 can be guaranteed.

The problem of number theoretical transforms is to determine such quantities M, a, N, and N^{-1} that the conditions

$$a^N \equiv 1 \qquad \mathit{modulo}\ M\ ,$$

$$(a^k - 1)^{-1} \quad \text{exists for} \quad 1 \le k < N\ ,$$

$$N^{-1} \quad \text{exists}$$

are fulfilled. The existence of such tupels (M, a, N, N^{-1}) has to be proved in each case. That is an essential difference to Fourier transform where for all values of N the same general expression $e^{2\pi i/N}$ can be used.

In table 5.3.-1, the minimal moduls M, the minimal values a, and the corresponding inverses N^{-1} are listed for given periods N. Only prime numbers M are used in this table. We can investigate also other numbers as the example $(M,a,N)=(9,2,6)$ with $2^6\equiv 1$ *modulo* 9 shows. But there is no inverse N^{-1} in this case.

In general, a solution can be guaranteed by Fermat's law if the greatest common divisor of M and a is $gcd(M,a)=1$. Then it follows

$$a^{\varphi(M)} \equiv 1 \qquad \mathit{modulo}\ M\ .$$

Here $\varphi(M)$ means the Eulerian divisor function (see section 4.3.2, [Kr81]).

Table 5.3.-1: Minimal prime moduls M and minimal values a for periods N

N	M_{min}	a	N^{-1}	a_{min}	M	N^{-1}
2	3	2	2	2	3	2
3	7	2	5	2	7	5
4	5	2	4	2	5	4
5	11	3	9	2	31	25
6	7	3	6	3	7	6 [1)]
7	29	7	25	2	127	109
8	17	2	15	2	17	15
9	19	4	17	2	73	65
10	11	2	10	2	11	10
11	23	2	21	2	23	21
12	13	2	12	2	13	12
13	53	10	49	2	8191	7561
14	29	4	27	2	43	40
15	31	7	29	2	151	141
16	17	3	16	2	257	241
17	103	8	97	2	131071	123361
18	19	2	18	2	19	18
19	191	5	181	2	524287	496693
...	...	...	...	...	....	....
58	59	2	58	2	59	58
59	709	20	697	2	179951	176901
60	61	2	60	2	61	60
...	...	...	...	...	....	....
64	193	11	190	2	641	631
128	257	9	255	5	769	763 [2)]
256	257	3	256	3	257	256 [3)]

1) there is no solution for $a_{min}=2$
2) there is no $a_{min}<5$ up to $M=10000$
3) there is no $a_{min}<3$ up to $M=10000$

Because we wish to work with gray values g within an interval $0 \le g \le g_{max}$, all results have to be also within this interval without overflow in respect of the modul M. Therefore M should be greater than g_{max} (see also section 5.3.4). Now, a period $N > M$ is impossible since only M different residue classes exist.

In practical applications, we have to search for such moduls M that $M > max(N, g_{max})$. In table 5.3.-2, some corresponding values M, N^{-1}, and a are given for $N=30$.

Table 5.3.-2: Moduls M and values a for $N=30$

M	N^{-1}	possible values a
31	30	3, 11, 12, 13, 17, 21, 22, 24
61	59	4, 5, 19, 36, 39, 45, 46, 49
77	75	2, 3, 5, 17, 18, 19, 24, 26, 30, 31, 38, 39, 40, 46, 47, 51, 52, 59, 61, 68, 72, 73, 74, 75
143	139	4, 17, 29, 30, 35, 36, 49, 61, 62, 68, 69, 74, 75, 82, 94, 95, 101, 107, 108, 114, 127, 134, 139, 140
151	146	23, 46, 66, 75, 113, 135, 147, 149

For $M > a^N$, no solution of $a^N \equiv 1$ *modulo* M is possible. For that reason, the modul M has to be fulfilled the requirements

$$N < M < a^N$$

for given values of a and N. Therefore we have no solution for $(N,a)=(6,2)$ because a systematic search shows that for all $M < 2^6 = 64$ no solution exists [Cr86]. The tupel $(N,a)=(6,2)$ is the only example for which the requirements of theorem 5.2.-3 cannot be fulfilled.

With respect to present computer constructions, powers of 2 have a great importance because the multiplication by a power of 2 can be performed by a shift of binary numbers stored in the computer memory. We can determine for any given periods N and roots $a=2^k$ minimal moduls M by a systematical search (table 5.3.-3).

Table 5.3.-3: Minimal moduls M for given periods N and roots $a=2^k$

$a=2^k=$ / $N=$	2	4	8	16	32	64	128
2	3	5	9	17	33	65	129
3	7	7	73	35	151	73	337
4	5	17	13	257	41	85	145
5	31	11	31	31	601	77	2201
6	[1]	13	19	119	77	95	301
7	127	43	127	29	71	127	[2]
8	17	257	17	[2]	51	97	255
9	73	19	[2]	19	73	[2]	511
10	11	41	11	187	251	123	213
11	23	23	23	23	89	67	2047
12	13	119	37	97	61	323	203
13	8191	2731	79	53	8191	79	911

[1] no solution
[2] no solution with $M<32767$

5.3.4 Two-Dimensional NTT

With respect to practical applications, the period N is the most important characteristics of a number theoretical transform. Therefore we have to search for moduls M for which N^{-1} at least one a exist.

This search has to be time effective. Now, the examples of table 5.3.-1 show that the chance to find an appropriate M is high when $M=k\cdot N+1$ is proved. In this case, the determination of N^{-1} is possible by

$$N^{-1} = M - \mathrm{ent}(M/N)$$

because of

$$N^{-1} \cdot N \equiv \left(k \cdot N + 1 - ent\left(\frac{k \cdot N + 1}{N}\right) \right) \cdot N \quad modulo \ (k \cdot N + 1)$$

$$\equiv k \cdot N + 1 - k \cdot N \equiv 1 \quad modulo \ (k \cdot N + 1) \ .$$

The existence of a root a is guaranted if $gcd(M,a)=1$, i.e. it is not necessary that M is a prime number (see $M=77=7 \cdot 11$ or $M=143=11 \cdot 13$ in table 5.3.-2). But it is necessary that $d=gcd(M,N)=1$ since

$$N \cdot x \equiv (k \cdot d) \cdot x \equiv (k \cdot x) \cdot d \quad modulo \ l \cdot d$$

$$\not\equiv 1 \quad modulo \ l \cdot d \ .$$

Therefore we will only investigate prime numbers $M=k \cdot N+1$.

For $N_1 \cdot N_2$-images, we have to work with only one modul M, i.e. we have to determine two roots a_1 and a_2 with

$$a_1^{N_1} \equiv 1 \quad modulo \quad M$$

$$a_2^{N_2} \equiv 1 \quad modulo \quad M \ .$$

This requirement is necessary because we need an unambiguously defined algebraic structure in which additions of any two elements are allowed, i.e. especially the addition of a_1 and a_2. If $S=scm(N_1,N_2)=N_1 \cdot N_2/gcd(N_1,N_2)$ is the *smallest common multiple* of N_1 and N_2, then $M=k \cdot S+1$ are moduls which have to be checked. Once more, M should be a prime number although there are other possibilities, too:

$$\begin{array}{llll} N_1 = 6 & M = 361 & N_1^{-1} = 301 & a_1 = 69 \\ N_2 = 9 & M = 361 & N_2^{-1} = 321 & a_2 = 28 \end{array}$$

or

$$\begin{array}{llll} N_1 = 4 & M = 561 & N_1^{-1} = 421 & a_1 = 89 \\ N_2 = 10 & M = 561 & N_2^{-1} = 505 & a_2 = 16 \ . \end{array}$$

The use of number theoretic transform in image processing is a fascinating method because of its integer properties and its computational efficiency. But there are difficulties for interpretation of the numerical results. Especially, there is no ordering relation like $3<5$ for integer numbers (see figure 5.3.-2), and therefore the multiplication of gray values in respect of a given modul does not yield any contrast enhancement of an image.

Suppose for instance that we work with $M=13$, $a=2$, and $N=12$. Given an image with gray values $g\in\{0,1,2,\ldots,12\}$. It is possible to multiply the gray values of all pixels by 5 using the following look-up-table:

g_{old}	0	1	2	3	4	5	6	7	8	9	10	11	12
g_{new}	0	5	10	2	7	12	4	9	1	6	11	3	8

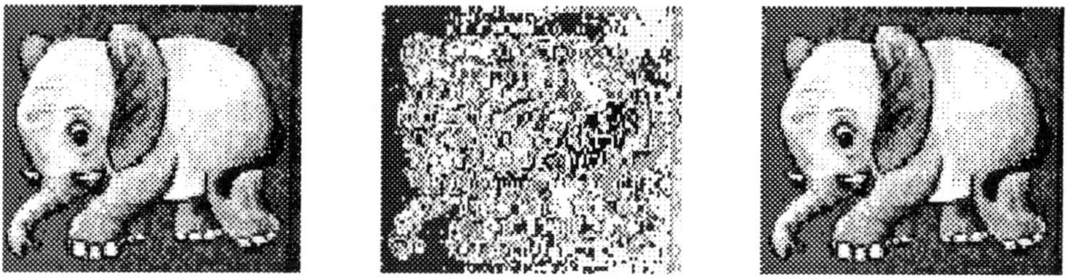

Fig. 5.3.-1: Multiplication of gray values without resulting contrast enhancement

In figure 5.3.-1, the left image contains 13 different gray values $g\in\{0,1,\ldots,12\}$. The second image contains the new gray values in respect of the above look-up-table. The gray value transformation $g_{\mathrm{new}}\equiv 5\cdot g_{\mathrm{old}}$ *modulo* 13 is a one-to-one transformation, but it is not possible to "see" any details in the new image.

However it is possible, to reconstruct the original image point by point using the relationship $5^{-1}\equiv 8$ *modulo* 13, i.e. the right image can be obtained with $g'\equiv 8\cdot g$ *modulo* 13.

Therefore, the main difference between the common Fourier transform and the number theoretical transform is that the quantities q *modulo* M cannot be arranged in a linear order as is possible for gray values used in practice.

Of course, such arrangement is likewise impossible for complex numbers. But we can compute the amplitudes $\sqrt{(c\cdot c^*)}$ of complex numbers c to have an arrangement rule, and we can arrange complex numbers with vanishing imaginary part, i.e. real valued results of image processing operations.

In figure 5.3.-2, the powers $\boldsymbol{a}^k$ with $\boldsymbol{a}^{13}=e$ and $a\equiv 2$ *modulo* 13 are circularly arranged like complex numbers $\exp(2\pi i/13)$ at the unit circle. We see that "conjugated" quantities $q=a^k$ and $q^*=a^{-k}$ fulfill always the formula $q\cdot q^*=e$. At the "unit circle", this definition of *conjugated numbers* is identical with the common definition of conjugated complex numbers.

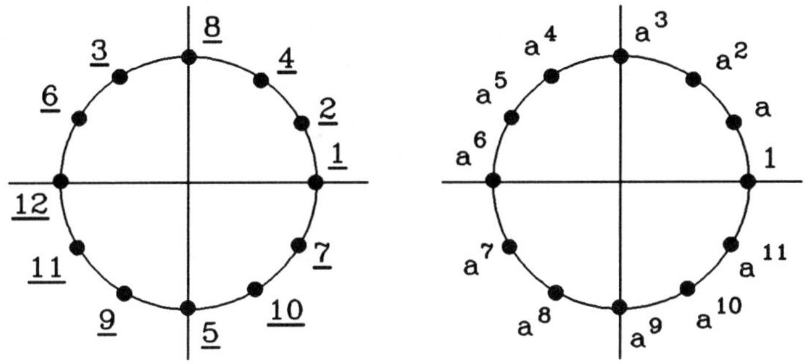

Fig. 5.3.-2: Circular arrangement of powers $\boldsymbol{a}^k$ with $N=12$

In case M is an odd prime number, i.e. $(M-1)/2=N/2$ exists, we will investigate the quantity

$$x = \mathbf{a}^l + \mathbf{a}^{l+N/2} \quad .$$

It is

$$x \cdot \mathbf{a}^{N/2} = \mathbf{a}^{l+N/2} + \mathbf{a}^l = x \ ,$$

and therefore $x(\boldsymbol{a}^{N/2}-1)=0$ follows. We have assumed that $\boldsymbol{a}^l \neq o$ for $0<l<N$ and that we work within a domain of integrity. Because of these assumptions, we obtain

$$x = \mathbf{a}^l\left(e + \mathbf{a}^{N/2}\right) = o \ ,$$

and therefore $\boldsymbol{a}^{N/2}=-e$.

Using number theoretical transforms, there is the problem of overflow in case of single convolutions $F=G*H$ because the finite field of residue classes are not an arranged field (there is no possibility to say that $n \cdot \boldsymbol{a}^k < m \cdot \boldsymbol{a}^k$ for $n<m$). But we can estimate as a necessary condition [Lu89] that

$$F_n \leq \max_n(F_n) \leq N \cdot \max_n(G_n) \cdot \max_n(H_n) \leq M-1 \quad .$$

Let $g_{\max}$ be the maximal gray value, i.e. $0 \leq F_n, G_n, H_n \leq g_{\max}$. Then the modul M must fulfill the condition

$$M > G_{\text{extr}} = g_{\max}^2 \cdot N \quad .$$

This condition is right for a single one-dimensional convolution where the result lies between 0 and G_{extr}. But if we have to work with multiple convolutions, say with $F_1*F_2*\dots*F_k$, then the requirement

$$M > G_{extr}^{(k)} = g_{max}^{k} \cdot N^{k-1}$$

has to be fulfilled.

In the special case of *Fermat transform* [Lu89] where

$$a = 2 \quad , \quad N = 2^{d+1} \quad , \quad M = F_d = 2^{2^d} + 1 \ ,$$

and F_d is the d-th *Fermat number*, the values g_{max} of table 5.3.-4 are allowed for k=2 and k=3.

Table 5.3.-4: Maximal possible values g_{max} for Fermat transforms

d	$N=2^{d+1}$	$M=F_d$	k=2 g_{max}	k=3 g_{max}
1	4	5	1	-
2	8	17	1	-
3	16	257	4	1
4	32	65537	45	4
5	64	$4.294\dots10^9$	≈7866	≈101

We can conclude that the number theoretical transforms are very interesting for the theory of discrete functions and that they are also useful for selected problems in image processing, but that the Fourier transform holds the most advantages with respect to all other transforms.

5.4 Inversion of Convolutions

5.4.1 Conditions for Inverse Elements

We use in this chapter the Fourier transform defined by

$$F_k = \frac{1}{N}\sum_{l=0}^{N-1} \varphi_l(F)\, e^{\frac{2\pi ikl}{N}} \quad , \quad \varphi_l(F) = \sum_{k=0}^{N-1} F_k\, e^{-\frac{2\pi ikl}{N}} \quad .$$

Further, the l-th Fourier coefficient of any function F shall be denoted by $\varphi_l(F)$. The unit functions $E^{(m)}$ introduced in section 5.1.2 have the Fourier transform

$$E_k^{(m)} = \frac{1}{N}\sum_{l=0}^{N-1} \varphi_l(E^{(m)})\, e^{\frac{2\pi ikl}{N}} = \delta_{m-k} \quad ,$$

$$\varphi_l(E^{(m)}) = \sum_{k=0}^{N-1} E_k^{(m)}\, e^{-\frac{2\pi ikl}{N}} = e^{-\frac{2\pi iml}{N}} \quad .$$

As shown in section 5.1.3, the convolution $H*F$ of two functions is equivalent to the application of an LSI-operator to function F, where the operator is determined by the circulant matrix C_H with

$$(C_H)_{i,k} = H_{i-k} \quad ,$$

and the function F is determined by the vector $(F_0,F_1,\ldots,F_{N-1})^T$:

$$C_H\, F = \begin{pmatrix} H_0 & H_{N-1} & H_{N-2} & \cdot & \cdot & \cdot & H_2 & H_1 \\ H_1 & H_0 & H_{N-1} & \cdot & \cdot & \cdot & H_3 & H_2 \\ H_2 & H_1 & H_0 & \cdot & \cdot & \cdot & H_4 & H_3 \\ \cdot & \cdot & \cdot & \cdot & \cdot & \cdot & \cdot & \cdot \\ \cdot & \cdot & \cdot & \cdot & \cdot & \cdot & \cdot & \cdot \\ H_{N-2} & H_{N-3} & H_{N-4} & \cdot & \cdot & \cdot & H_0 & H_{N-1} \\ H_{N-1} & H_{N-2} & H_{N-3} & \cdot & \cdot & \cdot & H_1 & H_0 \end{pmatrix} \begin{pmatrix} F_0 \\ F_1 \\ F_2 \\ \cdot \\ \cdot \\ F_{N-2} \\ F_{N-1} \end{pmatrix} \quad .$$

Every matrix in an N-dimensional vector space corresponds to a linear operator but not necessarily to a convolution operator. Assuming that a "convolution inverse function" H^{-1} corresponding to H exists, the convolution equation $H*H^{-1}=E^{(0)}$ is equivalent to the matrix equation $C_H \cdot C_H^{-1}=I$ where C_H is the circulant matrix corresponding to function H. Because C_H^{-1} corresponds to H^{-1} and $H^{-1}*H=E^{(0)}$, C_H^{-1} has to be also a circulant matrix.

For given functions H and G, a solution F of the equation $H*F=G$ only exists if the inverse matrix C_H^{-1} exists. This fact requires that the determinant of C_H is nonzero. Because the matrix C_H is a circulant matrix, we can formulate a criterion which guarantees the existence of inverse elements in respect of convolution:

<u>Theorem 5.4.-1:</u> The simple condition

$$\sum_{i=1}^{N-1} |H_i| < |H_0|$$

is sufficient that H^{-1} exists with regard to the operation $*$.

In connection to this theorem, the stepwise convolution iteration

$$F^{[m+1]} = \left(E^{(0)} - \frac{H}{H_0}\right) * F^{[m]} + \frac{G}{H_0}$$

with any start function $F^{[0]}$ generates a sequence of functions $F^{[m]}$ with

$$\lim_{m\to\infty} F^{[m]} = F \qquad \text{and} \qquad H*F=G \ .$$

If G is equal to the neutral element $E^{(0)}$, then the sequence converges to the inverse element H^{-1} of H.

To prove this theorem, we use the Banach fixpoint theorem for a so-called "contraction operator" [Ac75]. In the vector-space $\boldsymbol{F}$, we can use the function-norms

$$\|F\|_p = \left(\sum_{i=0}^{N-1} |F_i|^p\right)^{\frac{1}{p}}$$

for the functions $F \epsilon \boldsymbol{F}$ and any positive integer p. Then we have to work with an operator-norm corresponding to the function-norm which has to be the "compatible" operator-norm of the circulant matrix C_B corresponding to function

$$B = E^{(0)} - \frac{H}{H_0} \ .$$

This function B will be used because the equation $F=B*F+G/H_0$ is equivalent to the equation $F*H=G$. Practicable norms [Bu85] for numerical investigations are the compatible operator-norm

$$\|C_B\|_1 = \max_j \left(\sum_i |(C_B)_{i,j}|\right) ,$$

and the norm

$$\| C_B \|_\infty = \max_i \left(\sum_j |(C_B)_{i,j}| \right) .$$

Because C_B is a circulant matrix, it follows

$$\|C_B\|_1 = \|C_B\|_\infty .$$

Now, it is sufficient for the existence of C_B^{-1} that the norm of C_B is smaller than 1. Because we have for the 0-th coefficient

$$B_0 = \left(E^{(0)} - \frac{H}{H_0} \right)_0 = 0 ,$$

we obtain that

$$\|C_B\| = \sum_{i=1}^{N-1} \left| \frac{H_i}{H_0} \right| < 1 \quad \text{or} \quad \sum_{i=1}^{N-1} |H_i| < |H_0| .$$

Theorem 5.4.-2: Let H be any discrete periodical function. If and only if the condition

$$\max_i | \varphi_i(H) | < 1$$

is fulfilled for the Fourier coefficients of H, the convolution-iteration

$$F^{[m+1]} = H * F^{[m]} + G$$

using any start function $H^{[0]}$ converges always with

$$\lim_{m \to \infty} F^{[m]} = F \quad \text{and} \quad F = (E^{(0)} - H)^{-1} * G ,$$

and the inverse element $(E^{(0)} - H)^{-1}$ exists.

A necessary and sufficient convergence criterion [Bu85] for the iteration

$$F^{[m+1]} = C_H F^{[m]} + G$$

is that the eigen values λ_l of C_H satisfy the condition

$$\max_i |\lambda_i| < 1$$

because the eigenvalues λ_l of the circulant matrix C_H (see theorem 5.3.-2) are

$$\lambda_l(C_H) = \varphi_l(C_H) \quad .$$

5.4.2 Deconvolutions and Texture Synthesis

The investigations of section 5.4.-1 can be generalized also to two and more dimensions. In respect of theorem 5.4.-1, the sum of all absolute values of "non-central" elements of an n-dimensional discrete periodical function has to be smaller than the absolute value $|H_0|$ of the "central" element H_0. Therefore we have obtained a very simple method for deconvolution of images blurred by a function H.

In the case of image restoration described by $H*F=G$, the function H is the point spread function where H is mostly a small convolution kernel, i.e. only coefficients H_k with small values of $|k|$ are different from zero. Then, we can restore the original image F by a convolution iteration with the simple small kernel

$$B = \left(E^{(0)} - \frac{H}{H_0} \right)$$

if the condition of theorem 5.4.-1 is fulfilled. An example demonstrates this method: If a picture is blurred by the point spread function

$$H = \begin{bmatrix} .10 & .00 & .02 & .02 & .02 & .00 & .10 \\ .00 & .02 & .00 & .00 & .00 & .02 & .00 \\ .02 & .00 & .05 & .00 & .05 & .00 & .02 \\ .02 & .00 & .00 & \mathbf{1.00} & .00 & .00 & .02 \\ .02 & .00 & .05 & .00 & .05 & .00 & .02 \\ .00 & .02 & .00 & .00 & .00 & .02 & .00 \\ .10 & .00 & .02 & .02 & .02 & .00 & .10 \end{bmatrix}$$

with $H_0=1$ (the condition of theorem 5.4.-1 is fulfilled), we get with respect to successive applications of the small convolution kernel

$$B = \begin{bmatrix} -.10 & .00 & -.02 & -.02 & -.02 & .00 & -.10 \\ .00 & -.02 & .00 & .00 & .00 & -.02 & .00 \\ -.02 & .00 & -.05 & .00 & -.05 & .00 & -.02 \\ -.02 & .00 & .00 & \mathbf{0.00} & .00 & .00 & -.02 \\ -.02 & .00 & -.05 & .00 & -.05 & .00 & -.02 \\ .00 & -.02 & .00 & .00 & .00 & -.02 & .00 \\ -.10 & .00 & -.02 & -.02 & -.02 & .00 & -.10 \end{bmatrix}$$

a sequence of approximatively restored pictures. In figure 5.4.-1, the original picture F, the blurred picture $H*F$, and the restored picture $F^{[1]}=B*G+G$ are shown.

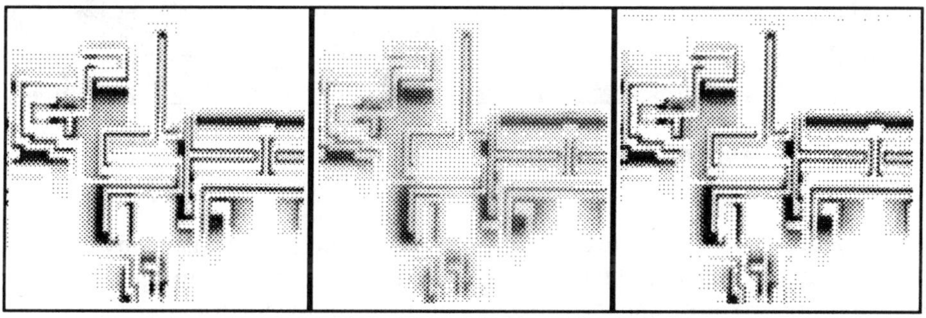

Figure 5.4.-1: Original picture (left), blurred picture, and restored picture (right) after one iteration

A further application of theorems 5.4.-1/2 is the generation of textures by an autoregressive process where the random variable X_p at lattice point p is given by a linear combination of random variables in a surrounding S_p of p and by an additive random process R_p [Ch82, Sc91a]:

$$X_p = \sum_{\substack{q \neq p \\ q \in S_p}} A_q X_{p-q} + R_p \ .$$

This equation can be represented as convolution

$$X = A*X + R \qquad , \qquad A_0=0$$

of a function X of random variables with any discrete periodical function A. The stochastic random field R is uncorrelated with expectation value $E(R_p)=k_1$ and dispersion $E(R_p^2)=k_2$ where k_1 and k_2 are constants. For the generation of a texture (i.e. any realization of the process), we can select such filter masks A that the conditions of theorems 5.4.-1/2 are satisfied.

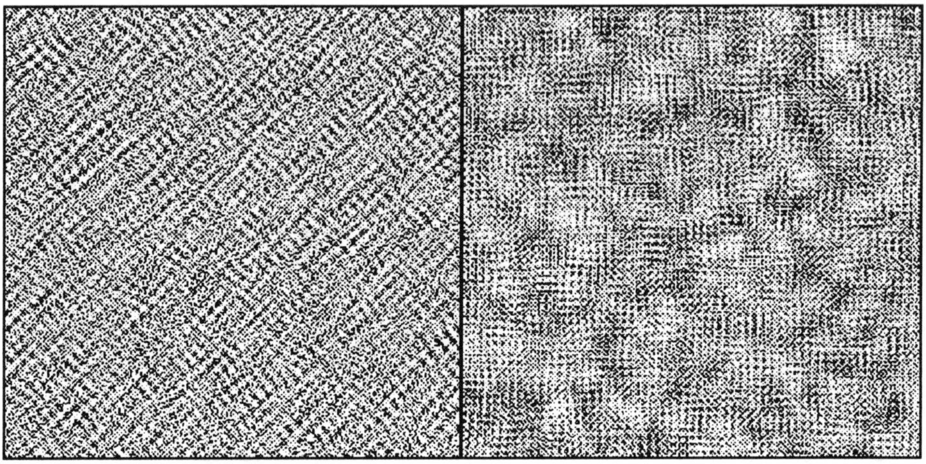

Figure 5.4.-2: Generation of textures after three iterations

In figure 5.4.-2, two textures are generated on the one hand by a two-dimensional asymmetric filter mask A_a with a very small kernel (generating a non-isotropic texture) and on the other hand by a symmetric filter mask A_s (generating a nearly isotropic texture):

$$A_a = \begin{bmatrix} -.10 & .00 & .00 & .00 & .35 \\ .00 & .00 & .00 & .00 & .00 \\ .00 & .00 & \mathbf{0.00} & .00 & .00 \\ .00 & .00 & .00 & .00 & .00 \\ .35 & .00 & .00 & .00 & -.10 \end{bmatrix}, \quad A_s = \begin{bmatrix} .22 & .00 & .00 & .00 & .2 \\ .00 & .00 & .00 & .00 & .0 \\ .00 & .00 & \mathbf{0.00} & .00 & .0 \\ .00 & .00 & .00 & .00 & .0 \\ .22 & .00 & .00 & .00 & .2 \end{bmatrix}$$

5.4.3 Approximative Computation of Inverse Elements

It cannot always be guaranteed that deconvolution is possible for a convolution $H*F=G$ because the determinant of the circulant matrix C_H is zero. But for such problems (i.e. the inverse H^{-1} and the inverse C_H^{-1} don't exist), we can define a solution, called "solution by pseudoinverse elements".

By means of the least-squares-method, we will now demonstrate the computation of "pseudoinverse elements" [Vo88a]. Let H be an LSI-filter. We want to compute a "pseudoinverse" filter H^+ such that $H*H^+$ yields approximatively the unit element $E^{(0)}$. For example, we want to compute the inverse element of the one-dimensional small kernel filter

$$H = \begin{bmatrix} \frac{1}{4} & \frac{2}{4} & \frac{1}{4} \end{bmatrix}$$

which corresponds to the discrete periodical function

$$H = \left(\frac{2}{4}, \frac{1}{4}, 0, \ldots, 0, \frac{1}{4} \right) .$$

We assume, that the "pseudoinverse filter" H^+ is likewise "small" compared with the number N of points of the one-dimensional image G, and that H^+ is symmetric like H:

$$H^+ = \begin{bmatrix} d & c & b & a & b & c & d \end{bmatrix} .$$

The corresponding function is $H^+ = (H_0^+, H_1^+, \ldots, H_{N-1}^+) = (a,b,c,d,0,\ldots, 0,d,c,b)$. The unknowns a,b,c,d are to determine.

The wanted convolution $H*H^+ = E^{(0)}$ yields the equation

$$\frac{1}{4} \begin{bmatrix} d \\ 2d+c \\ d+2c+b \\ c+2b+a \\ 2a+2b \\ a+2b+c \\ b+2c+d \\ c+2d \\ d \end{bmatrix} = \begin{bmatrix} 0 \\ 0 \\ 0 \\ 0 \\ 1 \\ 0 \\ 0 \\ 0 \\ 0 \end{bmatrix}$$

where all other coefficients vanish. But this system of nine equations cannot be satisfied.

Using the the least-squares-method, we obtain as sum over all squared deviations the expression

$$S(a,b,c,d) = 2d^2 + 2(2d+c)^2 + 2(d+2c+b)^2 + 2(c+2b+a)^2 + (2a+2b-4)^2 .$$

The requirement $S \to$ minimum yields $\partial S/\partial a = 0$, $\partial S/\partial b = 0$, $\partial S/\partial c = 0$, $\partial S/\partial d = 0$, and therefore we obtain the linear system of equations

$$\begin{pmatrix} 3 & 4 & 1 & 0 \\ 4 & 7 & 4 & 1 \\ 1 & 4 & 6 & 4 \\ 0 & 1 & 4 & 6 \end{pmatrix} \begin{pmatrix} a \\ b \\ c \\ d \end{pmatrix} = \begin{pmatrix} 4 \\ 4 \\ 0 \\ 0 \end{pmatrix} .$$

The solution $(a,b,c,d) = (40/9,-24/9,12/9,-4/9)$ yields the pseudoinverse small kernel filter

$$H^+ = \left[\frac{-4}{9}, \frac{12}{9}, \frac{-24}{9}, \frac{40}{9}, \frac{-24}{9}, \frac{12}{9}, \frac{-4}{9} \right] ,$$

and therefore we obtain approximatively the neutral element $E^{(0)}$:

$$H * H^+ = \left[\frac{-1}{9}, \frac{1}{9}, \frac{-1}{9}, \frac{1}{9}, \frac{8}{9}, \frac{1}{9}, \frac{-1}{9}, \frac{1}{9}, \frac{-1}{9} \right] .$$

Using this method, we can define approximative deconvolutions. The quality of the computed deconvolutions depends on the extent of the spatially-constrained "pseudoinverse" convolution kernel and of the strength with which the requirements of theorems 5.4.-1/2 are fulfilled.

5.4.4 Theory of Approximative Inversion

Now, we will investigate the general problem of the solution of convolution equation $F*X=E^{(0)}$. In common, the convolution theorem 5.2.-2 for Fourier transform is used to determine the function X:

$$\xi_k = \frac{\varepsilon_k^{(0)}}{\varphi_k} = \frac{1}{\varphi_k} , \qquad X_k = \frac{1}{N} \sum_{l=0}^{N-1} \frac{e^{\frac{2\pi ikl}{N}}}{\varphi_k} .$$

But here the problem arises that the solution is not defined if any $\varphi_k=0$ and that an existing solution is not locally restricted in general, respectively. But it is desirable for practice that we can work with locally restricted functions X.

Such a restricted function X will fulfill the given convolution equation only approximately, i.e. $F*X \approx E^{(0)}$. This requirement means for real valued functions that the scalar product

$$S = \langle F*X - E^{(0)} , F*X - E^{(0)} \rangle$$

reaches a minimal value. Supposing that X is an optimal solution, a small variation ΔX does not alterate the value of S, i.e. $S+\Delta S=S$ and $\Delta S=0$. We obtain

$$\Delta S = \langle F*X-E^{(0)}, F*\Delta X\rangle + \langle F*\Delta X, F*X-E^{(0)}\rangle + \langle \Delta X, \Delta X\rangle .$$

Neglecting the quadratic term $<\Delta X,\Delta X>$ and using formula 5.2.-1 for scalar products of convolutions, it follows

$$\Delta S = 2\langle (F*X-E^{(0)})*F^{\leftrightarrow}, \Delta X\rangle = 0$$

where $F^{\leftrightarrow}$ is the reflected function $(F^{\leftrightarrow})_k = F_{-k}$.

Taking into account that we can choose any non-vanishing coefficient values ΔX_k, it follows that the left side of this scalar product must be identical to zero:

$$((F*X-E^{(0)})*F^{\leftrightarrow})_q = 0 \quad \text{for} \quad q \in Q .$$

Here Q is the set of all points of the underlying translative neighborhood structure $[P,N]$ for which X can be different from zero, i.e. the locally restricted support of X. This last equation can be reformulated to

$$(F*F^{\leftrightarrow}*X)_q = (F^{\leftrightarrow})_q \quad \text{for} \quad q \in Q .$$

Finally, we obtain a linear equation system of order *card*(Q):

Formula 5.4.-1:

$$\sum_{p\in Q\subset P} (F*F^{\leftrightarrow})_{q-p} X_p = F_q^{\leftrightarrow} \quad \text{for} \quad q \in Q .$$

The obtained solution X of this equation system is the *pseudoinverse* of the original convolution equation $F*X=E^{(0)}$ which is not soluable in general (see [Bu85, Ku76]). Formula 5.4.-1 is a valuable tool for image restoration and other applications because the solution X is a small-kernel-approximation of F^{-1}, and therefore $X*G$ is an approximative restoration of the original image H, blurred by $F*H=G$.

5.4.5 Examples of Inverse Filters

To demonstrate the practicability of the formula 5.4.-1, we will search for the symmetric pseudoinverse X=(.... x_4 x_3 x_2 x_1 x_0 x_1 x_2 x_3 x_4 ...) of the one-dimensional filter F=(.... 0.1 0.2 0.4 0.2 0.1 ...). We obtain the following equation systems for the unknowns x_i:

$26x_0$	$+40x_1$	$+24x_2$	$+8x_3$	$+2x_4 + \dots$	$= 40$
$20x_0$	$+38x_1$	$+24x_2$	$+13x_3$	$+4x_4 + \dots$	$= 20$
$12x_0$	$+24x_1$	$+27x_2$	$+20x_3$	$+12x_4 + \dots$	$= 10$
$4x_0$	$+13x_1$	$+20x_2$	$+26x_3$	$+20x_4 + \dots$	$= 0$
x_0	$+4x_1$	$+12x_2$	$+20x_3$	$+26x_4 + \dots$	$= 0$
				...	....

In respect of the largest index k of unknowns, that means a filter size $2k+1$, we obtain the solutions x_i given in table 5.4.-1.

Table 5.4.-1: Solutions x_i dependent from size k of equation system

k	x_0	x_1	x_2	x_3	x_4	x_5	x_6
0	1.5385						
1	3.8298	-1.4894					
2	3.8251	-1.4754	-0.0182				
3	4.3182	-1.5907	-0.5331	0.5436			
4	4.5399	-1.7439	-0.6099	0.8671	-0.2919		
5	4.5411	-1.7345	-0.6292	0.8633	-0.2468	-0.0451	
6	4.5877	-1.7491	-0.6802	0.9351	-0.2368	-0.1741	0.1255
7	4.5979	-1.7572	-0.6822	0.9553	-0.2619	-0.1796	0.1760
8	4.5986	-1.7564	-0.6852	0.9557	-0.2551	-0.1892	0.1742
15	4.6022	-1.7579	-0.6886	0.9623	-0.2565	-0.2011	0.1921

At the latest at a size of $k=15$, stability is reached. The solution for $k=15$ equals the solution of the inversion C_F^{-1} of the circulant convolution matrix C_F of the filter $F=(\dots.\ 0.1\ 0.2\ 0.4\ 0.2\ 0.1\ \dots)$, i.e. it is equal to the solution yielded by means of Fourier transform. But then, C_F has to be at least a $25 \cdot 25$-matrix.

The application of locally restricted (spatially-constrained) inverse filters is not necessary one-dimensional. The two-dimensional generalization is straightforward. The size of the equation system corresponding to a two-dimensional discrete periodical function can be very large, and therefore the numerical effort is high to compute the pseudoinverse. However for locally restricted $(2N+1)^2$-filters with octand-symmetry, we have only to compute the values of $(N+1)(N+2)/2$ variables.

In figure 5.4.-3, an original image was blurred by the following symmetric 7·7-filter F (only one octand is represented) for which the condition of theorem 5.4.-1 is not fulfilled:

0.16	0.04	0.02	0.01
	0.03	0.02	0.01
		0.02	0.01
			0.01

To compute a locally restricted pseudoinverse, we use a symmetric 11·11-filter X, and because of the assumed symmetry, we have only to determine as (computer aided) solution of formula 5.4.-1 the values of 21 variables instead of 121 variables:

7.952	-0.891	-0.038	-0.001	0.242	-0.027
	-0.377	-0.173	-0.064	0.178	-0.047
		-0.422	-0.077	0.169	-0.054
			-0.334	0.162	-0.036
				0.055	-0.008
					0.003

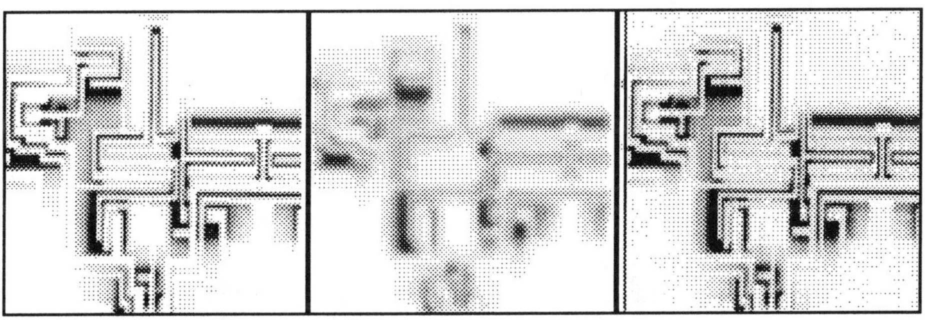

Figure 5.4.-3: Original picture (left), blurred picture (7·7-filter) and restored picture obtained using an 11·11-filter (right)

Also for filters in other lattices, we can compute pseudoinverse solutions [Vo88a]. Given the lattice $\Gamma(\lambda,\nu)=\Gamma(3,6)$ and the filter F shown on the left side of figure 5.4.-4. The points not characterized by coefficient values have zero-values. We use for the full-symmetric pseudoinverse X the coefficients $x_0,\ldots,x_5$ (right side of figure 5.4.-4). The solution of formula 5.4.-1 yields for the coefficients of X and for the coefficients of $F*X$ the values of table 5.4.-2.

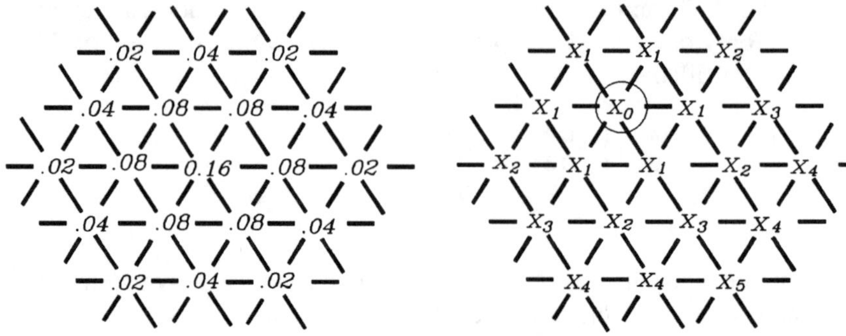

Fig. 5.4.-4: Locally restricted filter in $\Gamma(\lambda,\nu)=\Gamma(3,6)$

Table 5.4.-2: Coefficients of filter kernels X and F^*X in $\Gamma(3,6)$

k	X_k	$(F^*X)_k$
0	15.251	0.905
1	-3.081	0.010
2	-1.065	0.031
3	1.659	0.016
4	0.380	0.046
5	-0.607	-0.035

This example shows once more the advantages of the here derived *small-kernel method* of deconvolution because there are no other obvious and simple approaches for filter inversion in non-orthogonal lattices [Du84, Ja85].

5.5 Differences and Sums of Functions

5.5.1 Differences of One-Dimensional Discrete Functions

Real functions of real variables are the main topic of "higher mathematics", i.e. of analysis. Calculus, with differentiation and integration as basic operations for real functions, has a long successful history since the century of Newton and Leibniz.

We can consider differentiation and integration also from an algebraical standpoint in the sense that differential operator $\boldsymbol{L}_\partial$ and integral operator $\boldsymbol{L}_\int$ are linear operators in a given function space. For instance, the application of $\boldsymbol{L}_\partial$ to products $F \cdot G$ is defined by the following equation:

$$\boldsymbol{L}_\partial (F \cdot G) = (\boldsymbol{L}_\partial F) \cdot G + F \cdot (\boldsymbol{L}_\partial G) \quad .$$

Because of the practical importance of differentiation and integration for real functions, it is desirable to define corresponding notions in discrete spaces, too. Instead of differentiation, we will investigate differences of functions, and integration must be replaced by summation [Nö24, Me59, Be79, Sp82, St87].

Def. 5.5.-1: Let F be a one-dimensional discrete function with $F(k)=F_k$ as coefficients. We denote

$$\Delta^{(m)} F_k = \Delta^{(m-1)} F_{k+1} - \Delta^{(m-1)} F_k$$

$$\Delta^{(0)} F_k = F_k \quad , \quad \Delta^{(1)} F_k = \Delta F_k = F_{k+1} - F_k$$

as *differences* of k-th order.

The *difference operators* $\Delta^{(m)}$ are linear operators in the function space $\boldsymbol{F}$. The application to products of functions yields

$$\Delta (FG)_n = F_{n+1} G_{n+1} - F_n G_n = (\Delta F_n) G_{n+1} + F_n (\Delta G_n) \quad .$$

Functions F in discrete spaces can be given by three different methods:

- The coefficients F_k are determined by tables, lists, or data arrays as it is usual in image processing.

- The coefficients F_k are determined by functions of k as usual for real functions where $f(x)$ can be calculated using only the value x. In image processing, this method is used for description of basic functions.

- The coefficients F_k are determined by coefficients of other functions. This method is typical for image arithmetic (point-wise addition or subtraction of images) or local filtering (gradient operator, Laplace operator, median filtering etc.).

In this chapter, we will investigate functions which are defined by the second method (see, for instance, table 5.5.-1) because the laws and properties of such functions are not sufficiently known in image processing.

Table 5.5.-1: Differences $\Delta^{(m)}$ of function k^4

k	k^4	$\Delta^{(1)}$	$\Delta^{(2)}$	$\Delta^{(3)}$	$\Delta^{(4)}$
0	0	1	14	36	24
1	1	15	50	60	24
2	16	65	110	84	24
3	81	175	194	108	24
4	256	369	302	132	...
5	625	671	434	...	...
6	1296	1105	...	...	...
7	2401	...	...	...	...

As an example, we will investigate the polynomial function k^l, i.e. a polynomial of l-th order. It is

$$\begin{aligned} \Delta^{(m)} k^l &= \Delta^{(m-1)} (k+1)^l - \Delta^{(m-1)} k^l \\ &= \Delta^{(m-1)} \sum_{j=0}^{l} \binom{l}{j} k^j - \Delta^{(m-1)} k^l \\ &= \Delta^{(m-1)} \sum_{j=0}^{l-1} \binom{l}{j} k^j \; . \end{aligned}$$

Therefore, the m-th difference of a polynomial of l-th order consists of a sum of $(m-1)$-th differences of polynomials of j-th order with $j<l$. We find that the l-th difference of an l-polynomial is a constant function:

$$\Delta^{(l)} k^l = l! \; .$$

For binomials, the differences are [St87]

$$\Delta \binom{l}{j} = \binom{l+1}{j} - \binom{l}{j} = \binom{l}{j-1} \; ,$$

$$\Delta^{(2)} \binom{l}{j} = \Delta \binom{l+1}{j} - \Delta \binom{l}{j} = \binom{l+1}{j-1} - \binom{l}{j-1} = \binom{l}{j-2} \; ,$$

$$\Delta^{(m)} \binom{l}{j} = \binom{l}{j-m} \; .$$

In section 5.5.3, we will define the so-called *factorials* $k^{[l]}$ as special discrete functions

$$\begin{aligned}
k^{[1]} &= k \\
k^{[2]} &= k^2 - k \\
k^{[3]} &= k^3 - 3k^2 + 2k \\
k^{[4]} &= k^4 - 6k^3 + 11k^2 - 6k \\
k^{[5]} &= k^5 - 10k^4 + 35k^3 - 50k^2 + 24k
\end{aligned}$$

with

$$\begin{aligned}
k &= k^{[1]} \\
k^2 &= k^{[2]} + k^{[1]} \\
k^3 &= k^{[3]} + 3k^{[2]} + k^{[1]} \\
k^4 &= k^{[4]} + 7k^{[3]} + 6k^{[2]} + k^{[1]} \\
k^5 &= k^{[5]} + 15k^{[4]} + 25k^{[3]} + 10k^{[2]} + k^{[1]} \quad .
\end{aligned}$$

These functions $k^{[l]}$ fulfill the relationship

$$\Delta\, k^{[l]} = l \cdot k^{[l-1]} \quad .$$

5.5.2 Difference Equations and Z-Transform

A linear *difference equation* with constant coefficients is defined by

$$\sum_{j=0}^{l} a_j F_{k+j} = a_0 F_k + a_1 F_{k+1} + \ldots + a_l F_{k+l} = G_k \quad .$$

The discrete function G is assumed as known, and the unknown discrete function F has to be found. This equation is a *recursion formula* because with

$$F_{k+l} = \frac{1}{a_l}\left[G_k - \sum_{j=0}^{l-1} a_j F_{k+j} \right]$$

and knowledge about starting values $F_0,\ldots,F_{l-1}$ and function G all following values F_m with $m > l-1$ can be determined. For $G \equiv 0$, the equation is denoted as *homogeneous difference equation.* It shall be mentioned as interesting fact that such a difference equation can be described as convolution

$$\sum_{j=0}^{l} a_j F_{k+j} = \sum_{j=0}^{l} a_j f^{\leftrightarrow}_{-k-j} = (a * F^{\leftrightarrow})_{-k} = G_k \quad .$$

Thereby we have assumed that all coefficients a_j vanish but the coefficients $a_0 \ldots a_l$. Because of this, linear difference equations of periodical functions can be solved approximately with the method developed in chapter 5.4.

In contrast to that method, linear difference equations for nonperiodical discrete functions are treated with the Z-transform [St87]. Each function $F=(F_0, F_1, \ldots)$ defined for non-negative arguments shall be transformed to a function

$$\mathbf{Z}(F) = \Phi(z) = \sum_{k=0}^{\infty} F_k z^{-k} .$$

If the infinite sum exists, the function F is denoted as *Z-transformable.*

For image processing, it would be meaningful to work with "finite Z-transforms" for which the assumption of transformability is self-evident. Therefore, we could define

$$\Phi(z) = \sum_{k=0}^{N-1} F_k z^{-k}$$

as Z-transform of function F with periodicity N. Then, the important convolution operation yields

$$\mathbf{Z}(F*G) = \sum_{k=0}^{N-1} (F*G)_k z^{-k} = \sum_{k=0}^{N-1} \sum_{l=0}^{N-1} F_l G_{k-l} z^{-k}$$

$$= \sum_{l=0}^{N-1} F_l z^{-l} \sum_{k=0}^{N-1} G_{k-l} z^{-(k-l)} .$$

Now, if z^{-k} would be also a function with periodicity N, the main convolution equation $\mathbf{Z}(F*G)=\mathbf{Z}(F)*\mathbf{Z}(G)$ could be derived. But this requirement leads to the relationship $z^{-N}=z^N=e=1$ for the algebraic quantity z, i.e. to the main assumption of the theory of discrete periodical functions (see section 5.2.4 and chapter 5.3).

5.5.3 Sums of Functions

We denote the reverse operation of difference operation as *summation*. The corresponding task is to find a function G so that the differences of this function are equal to a given F, i.e.

$$\Delta G_k = F_k .$$

This problem is equivalent to searching for a solution of the very simple difference equation

$$G_{k+1} - G_k = F_k \ .$$

Assuming that G is a solution of this equation, we find

$$\sum_{k=k_0}^{l} F_k = \sum_{k=k_0}^{l} \left(G_{k+1} - G_k\right) = G_{l+1} - G_{k_0} \ .$$

Therefore, we have shown that linear difference equations, convolutions of discrete functions, and summations of functions are connected with one another. In general, there are the following properties of solutions of the above difference equation:

- If G_k is any solution of $\Delta G_k = F_k$, then $G_k + c$ is also a solution with c as a constant.
- If F_k is any polynomial of order l, then G_k as general solution of $\Delta G_k = F_k$ is a polynomial of order $l+1$.
- The Abelian theorem

$$\sum_{k=1}^{l} F_k G_k = F_{l+1} \sum_{k=1}^{l} G_k - \sum_{k=1}^{l} \Delta F_k \sum_{m=1}^{k} G_m$$

  is fulfilled [Sp82].

For summation of functions, the so-called *generalized powers* or *factorials* $k^{[l]}$ plays an important role:

$$k^{[l]} = k(k-1)\ldots(k-l+1) = \begin{cases} \dfrac{k!}{(k-l)!} & \text{for} \quad l \le k \\ 0 & \text{for} \quad l > k \end{cases} \ .$$

Here, l is a non-negative integer, $k^{[0]}=1$ and $k^{[1]}=k$ (see section 5.5.1). As most important property of the generalized power, we obtain

$$\Delta k^{[l]} = (k+1)^{[l]} - k^{[l]} = \frac{(k+1)!}{(k+1-l)!} - \frac{k!}{(k-l)!}$$

$$= \left(\frac{k+1}{k+1-l} - 1\right) \frac{k!}{(k-l)!} = \frac{l\,k!}{(k-(l-1))!}$$

or

$$\Delta k^{[l]} = l\, k^{[l-1]} .$$

Therefore

$$G_k = \frac{k^{[l+1]}}{l+1}$$

is a solution of the difference equation $\Delta G_k = k^{[l]}$. Because of that, the solution of the corresponding summation tasks is

$$\sum_{k=k_1}^{k_2} k^{[l]} = \left.\frac{k^{[l+1]}}{l+1}\right|_{k=k_1}^{k_2} = \frac{k_2^{[l+1]} - k_1^{[l+1]}}{l+1} .$$

Using this equation, the summation problem is solved for any polynomial P_k of degree m, too. One writes P_k as a series

$$P_k = \sum_{l=0}^{m} \frac{c_l}{l!}\, k^{[l]}$$

of generalized powers weighted by coefficients c_l, and it follows

$$\sum_{k=k_1}^{k_2} P_k = \sum_{l=0}^{m} \frac{c_l}{l!} \sum_{k=k_1}^{k_2} k^{[l]} = \sum_{l=0}^{m} \frac{c_l}{l!} \cdot \frac{k_2^{[l+1]} - k_1^{[l+1]}}{l+1} .$$

5.5.4 Bernoulli's Polynomials

Bernoulli's Polynomials are special polynomials [Nö24, Me59, Sp82] useful for investigation of differences and sums of functions, respectively. They are solutions of the difference equation

$$\Delta B_k^{(l)} = B_{k+1}^{(l)} - B_k^{(l)} = l\, k^{l-1}$$

(similar to the difference equation $\Delta k^{[l]} = l \cdot k^{[l-1]}$ for factorials). For the unambiguous definition of the solutions, the values

$$B_0^{(l)} = \beta_l$$

are given where the β_l are Bernoulli's numbers:

$$\beta_0 = 1 \qquad \beta_1 = -\frac{1}{2} \qquad \beta_2 = \frac{1}{6} \qquad \beta_3 = 0$$

$$\beta_4 = -\frac{1}{30} \qquad \beta_5 = 0 \qquad \beta_6 = \frac{1}{42} \qquad \beta_7 = 0$$

$$\beta_8 = -\frac{1}{30} \qquad \beta_9 = 0 \qquad \beta_{10} = \frac{5}{66} \qquad \beta_{11} = 0$$

$$\beta_{12} = -\frac{691}{2730} \qquad \beta_{13} = 0 \qquad \beta_{14} = \frac{7}{6} \qquad \beta_{15} = 0$$

There is an equation

$$\sum_{l=0}^{m} \binom{m+1}{l} \beta_l = 0$$

for Bernoulli's numbers from which with $\beta_0 = 1$ the recursion formula

$$\beta_m = -\frac{1}{m+1} \sum_{l=0}^{m-1} \binom{m+1}{l} \beta_l$$

follows. This formula is a special case of the recursion formula

$$B_k^{(m)} = (m+1)\, k^m - \sum_{l=0}^{m-1} \binom{m+1}{l} B_k^{(l)}$$

for Bernoulli's polynomials with

$$B_k^{(0)} = 1$$

for the 0-th polynomial. Therefore, we obtain

$$B_k^{(m)} = \sum_{l=0}^{m} \binom{m}{l} \beta_l\, k^{m-l} \quad .$$

The first polynomials are

$$\begin{aligned}
B_k^{(0)} &= 1 \;, \\
B_k^{(1)} &= k - \frac{1}{2} \;, \\
B_k^{(2)} &= k^2 - k + \frac{1}{6} \;, \\
B_k^{(3)} &= k^3 - \frac{3}{2} k^2 + \frac{1}{2} k \;, \\
B_k^{(4)} &= k^4 - 2k^3 + k^2 - \frac{1}{30} \;.
\end{aligned}$$

5.5.5 Determination of Moments

In image processing, sometimes there is the task to determine the value of *moments* m_{pq} of binary objects B, i.e.

$$m_{pq} = \sum_x \sum_y x^p y^q = \sum_y y^q \sum_{\substack{(x,y) \text{ with} \\ y=\text{const}}} x^p .$$

For y=const, we obtain single segments S_i of points of the binary object (figure 5.5.-1).

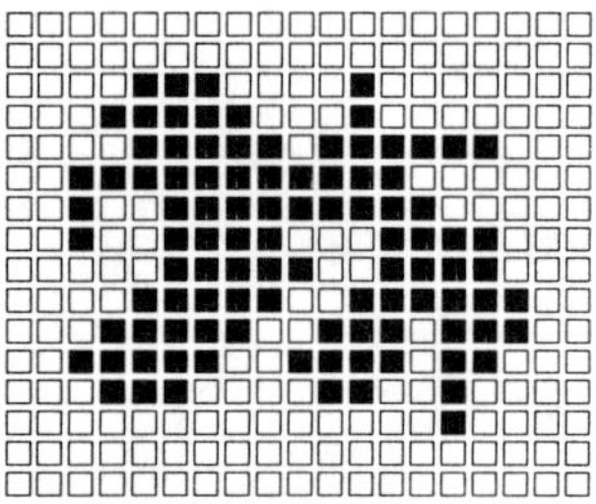

Fig. 5.5.-1: Binary object with point segments

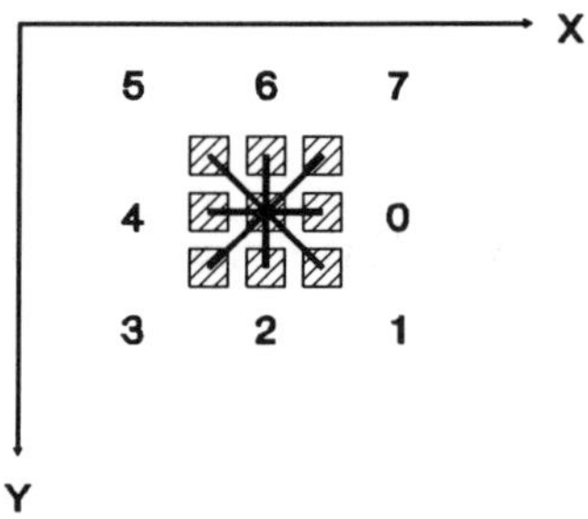

Fig. 5.5.-2: Direction code

Therefore, it is

$$m_{pq} = \sum_y y^q \sum_{S_i(y)} \sum_{x \in S_i(y)} x^p$$

whereby the sums are taken for all values y covered by the object, all segments $S_i(y)$, and all points (x,y) with $x \in S_i(y)$. We obtain for single segments the sum equation

$$\sum_{x \in S_i(y)} x^p = \sum_{x=xmin_i}^{xmax_i} x^p = \sum_{x=1}^{xmax_i} x^p - \sum_{x=1}^{xmin_i - 1} x^p$$

with $xmin_i$ und $xmax_i$ as extremal coordinate values of the segment S_i. Finally, it is sufficient to determine power sums of the type

$$Q_p(l) = \sum_{k=0}^{l} k^p$$

as a typical example for the problems investigated in section 5.5.3.

For the powers up to the fourth order, we have

$$Q_0(l) = \sum_{k=1}^{l} k^0 = l ,$$

$$Q_1(l) = \sum_{k=1}^{l} k^1 = \frac{l(l+1)}{2} ,$$

$$Q_2(l) = \sum_{k=1}^{l} k^2 = \frac{l(l+1)(2l+1)}{6} ,$$

$$Q_3(l) = \sum_{k=1}^{l} k^3 = \frac{l^2(l+1)^2}{4} ,$$

$$Q_4(l) = \sum_{k=1}^{l} k^4 = \frac{l(l+1)(2l+1)(3l^2+3l-1)}{30} .$$

Because of that, it is

$$m_{pq} = \sum_{y} y^q \sum_{S_i(y)} \left(Q_p(xmax_i) - Q_p(xmin_i-1)\right) .$$

The set of segment end points is a subset of the set of contur points of the binary object. Because of that, to determine the moments algorithmically we have to run through the contur and to decide for each contur point (x_i,y_i) whether it is a minimal or a maximal end point of a segment. This decision can be done using the direction d_{i-1} going to (x_i,y_i) and the direction d_i coming from (x_i,y_i).

In table 5.5.-2, it is represented whether p_i is an input point (I) or an output point (O) in respect of the object contur. For configurations designated by C no changes occur because in these points we have an input and an output (see figure 5.5.-3).

For conturs of holes, the corresponding formula is

$$\sum_{x\in S_i(y)} x^p = \sum_{x=xmin_i+1}^{xmax_i-1} x^p = \sum_{x=1}^{xmax_i-1} x^p - \sum_{x=1}^{xmin_i} x^p$$

$$= Q_p(xmax_i-1) - Q_p(xmin_i)$$

with $xmin_i$ and $xmax_i$ as contur point coordinates determining the hole contur (see [Vo92]).

Table 5.5.-2: Contur point characterization by contur direction

	$d_i=$ 0	1	2	3	4	5	6	7
d_{i-1} = 0		O	O	O	O			
1		O	O	O	O			
2		O	O	O	O	C	C	
3		O	O	O	O	C	C	
4	I					I	I	I
5	I					I	I	I
6	I	C	C			I	I	I
7	I	C	C			I	I	I

Fig. 5.5.-3: Input/output points of a contur

Some time effective methods of moment determination have already been developed in the past [Za87, St90, Ji91]. But the approach given here is explicitly directed to lattice points so that no approximations are necessary (see also [Ag85, Sa91] where this method is used for single segments).

Further, this approach allows a direct generalization to moments of objects in higher dimensional spaces Z^n. To see that, we refer to section 3.1.3 where we have investigated the surface detection of any n-dimensional objects.

If we are detecting an $(n-1)$-dimensional surface of an n-dimensional object B using the neighborhood algorithm described in section 3.1.3, all founded boundary pairs (p,q) with $p \in B$ and $q \notin B$ determine unambiguously one coordinate direction out of $2n$ possible directions. We can use one of the coordinates (for instance, say x_1) as *segment coordinate* so that a moment is given by

$$m_{p_1 \ldots p_n}(B) = \sum_{x \in B} x_1^{p_1} x_2^{p_2} \ldots x_n^{p_n} = \sum_{x_2, \ldots, x_n} x_2^{p_2} \ldots x_n^{p_n} \sum_{S_i} \sum_{x_1 = xmin_i}^{xmax_i} x_1^{p_1}$$

with $x=(x_1,x_2,\ldots,x_n)$ and $S_i=S_i(x_2,\ldots,x_n)$ as i-th segment determined by constant values of $x_2,\ldots,x_n$.

The open question is whether a surface point p is an input point or an output point (see figure 5.5.-3). In Z^n, we have to use a different criterion to that in Z^2: If we have $x_1(p)>x_1(q)$ for the choosed x_1-coordinate values of a boundary pair (p,q) with $p \in B$ and $q \notin B$, then p is an input point (minimal segment point) for the corresponding segment, and we obtain $xmin=x_1(p)$. In the other case of $x_1(p)<x_1(q)$, maximal segment points are determined by $xmax=x_1(p)$.

Because it is hardly possible to determine the corresponding minimal and maximal point for a given segment, we calculate the moments in respect of a $(n-1)$-dimensional surface S of the object B by rearrangement of the multiple sums:

$$m_{p_1 \ldots p_n}(S) = \sum_{\substack{(p,q) \\ x_1(p)<x_1(q)}} x_2^{p_2} \ldots x_n^{p_n} Q_{p_1}(x_1(p)) - \sum_{\substack{(p,q) \\ x_1(p)>x_1(q)}} x_2^{p_2} \ldots x_n^{p_n} Q_{p_1}(x_1(p)-1)) \ .$$

Here, the sums are to calculate for all boundary pairs (p,q) of the investigated surface S, and the Q's are power sums as given above.

The moment $m_{0\ldots0}$ is the n-dimensional "volume" V_i included into the i-th surface, i.e. the number of lattice points included into the surface or lying on the surface. If this volume is non-negative, the surface is an *outer surface* of the investigated object. In the other case, the surface is an *inner surface* and includes a *hole* of the object. If we are interested in the moments of such a hole, we have to use the formula

$$m_{p_1 \ldots p_n}(S) = \sum_{\substack{(p,q) \\ x_1(p)>x_1(q)}} x_2^{p_2} \ldots x_n^{p_n} Q_{p_1}(x_1(p)-1) - \sum_{\substack{(p,q) \\ x_1(p)<x_1(q)}} x_2^{p_2} \ldots x_n^{p_n} Q_{p_1}(x_1(p))) \ .$$

5.5.6 Final Comments

To complete the long path from the Königsberg bridges to the moments of grid point regions, we should here summarize the fundamental ideas which form the framework of this book. I have tried to show those mathematical results which are of interest or which could be of interest for digital image processing. These results are derived from set theory and graph theory, the theory of relations and the theory of algebraical structures, combinatorics and geometry. And they can all be categorized in the field of discrete mathematics.

The character of digital image processing demands that some mathematical investigations be explicitly depicted here. These investigations include above all the theory of homogeneous incidence structures with the important results of the matching formulas, searches in oriented neighborhood structures, curvatures of toroidal nets, similiarity of discrete objects, and connected object's characteristics. Furthermore, in the closed or open number lattice Z^n which can be regarded as a special incidence structure, new topological and geometric laws can be formulated (three-dimensional surface detection, intersection formulas, digitalization effects, maximal convex objects).

Finally, the closed number lattice Z^n can be employed as a support for discrete periodical functions. Convolution has proved to be the most important operation between such functions, and it provides a practical method for approximative deconvolutions by means of locally restricted filters. Summation of discrete functions as counterpart to analytical integration allows the time-effective calculation of moments of lattice point regions - another example of practical value. Theoretical investigations and methods prepared for practical purposes are presented here for the first time in a to some extent unified way.

In my opinion, the possibilities of discrete mathematics for digital image processing have until now been insufficiently exploited. At the same time, I would like to particularly emphasize the word "discrete" because often methods of analysis are too uncritically used whose results are employed for digital image interpretation only after (mostly uncontrollable) rounding. In contrast to this, I would hope to put forward the evidence that consistent discrete modelling of a digital image can lead to exact theoretical insights and to new time-effective algorithms.

And I would like to point out, at the risk of repetition, that digital images, as they exist in computer memory, are discrete data sets - i.e. there is no possibility of discrete storage of "continuous" point sets. The data sets of informatics in general, and of image processing in particular, are discrete both with regard to storage (integer addresses) and with regard to the gray values to be stored. The data of a stored digital image are a reality for whose description suitable models must be developed. This book would hope to be a contribution to this end.

6 Summary and Symbols

In this part of the book, the most important formulas and theorems shall be summarized so that the reader may have a comprehensive survey of the content dealt within the book.

Theorem 1.1.-1 (node theorem, vertex theorem): In each undirected graph $[P,N]$, the number of edges is given by $\varkappa = card(N)/2$, and $\varepsilon = card(P)$ is the number of vertices. Then

$$\sum_{p \in P} \nu(p) = 2\varkappa \qquad \text{and} \qquad \bar{\nu} = \frac{2\varkappa}{\varepsilon}$$

are fulfilled with $\bar{\nu}$ as mean neighborhood degree.

Theorem 1.3.-1 (complement theorem): The complementary set $\overline{C_i}$ of a single complementary component C_i of a point set $M \subset P$ of the neighborhood structure $[P,N]$ is always connected.

Formula 1.4.-1 (erosion and dilatation):

$$\boldsymbol{CEC} = \boldsymbol{D} \quad , \quad \boldsymbol{CDC} = \boldsymbol{E}$$

$$\boldsymbol{EC} = \boldsymbol{CD} \quad , \quad \boldsymbol{CE} = \boldsymbol{DC} \quad .$$

Formula 1.4.-2 (opening and closing):

$$\boldsymbol{CO} = \boldsymbol{SC} \quad , \quad \boldsymbol{OC} = \boldsymbol{CS} \quad ,$$

$$\boldsymbol{SE} = \boldsymbol{E} = \boldsymbol{EO} \quad ,$$

$$\boldsymbol{DS} = \boldsymbol{D} = \boldsymbol{OD} \quad ,$$

$$\boldsymbol{OO} = \boldsymbol{O} \quad , \quad \boldsymbol{SS} = \boldsymbol{S} \quad .$$

Formula 2.1.-1 (matching formulas):

$$a_k b_{kl} = a_l b_{lk} \qquad \text{for} \qquad 0 \le k, l \le n \; ,$$

$$\frac{b_{kl}\, b_{lm}}{b_{ml}\, b_{lk}} = \frac{b_{km}}{b_{mk}} \qquad \text{for} \qquad 0 \le k, l, m \le n \; .$$

Formula 2.1.-2 (structure constants):

$$b^{(n)}_{kl} = \begin{cases} 2^{l-k}\binom{n-k}{n-l} & \text{for } k<l \\ 1 & \text{for } k=l \\ 2^{k-l}\binom{k}{l} & \text{for } k>l \end{cases}$$

Formula 2.1.-3 (structure constants):

$$\frac{b_{lk}}{b_{kl}} = \frac{2^{l-k}\binom{l}{k}}{2^{l-k}\binom{n-k}{n-l}} = \frac{\binom{n}{k}}{\binom{n}{l}} \qquad \text{for} \quad k\le l \ ,$$

$$\frac{b_{lk}}{b_{kl}} = \frac{2^{k-l}\binom{n-l}{n-k}}{2^{k-l}\binom{k}{l}} = \frac{\binom{n}{k}}{\binom{n}{l}} \qquad \text{for} \quad k\ge l \ .$$

Formula 2.1.-4 (structure constants):

$$\sum_{k=0}^{n} (-1)^k \frac{b_{lk}}{b_{kl}} = \frac{1}{\binom{n}{l}} \sum_{k=0}^{n} (-1)^k \binom{n}{k} = 0$$

Theorem 2.2.-1 (mesh theorem): In an oriented neighborhood structure $[P,N,C]$, ω is the number of all different directed ways (meshes). Then

$$\sum_{m\in M} \lambda(m) = 2\varkappa \qquad \text{and} \qquad \bar{\lambda} = \frac{2\varkappa}{\omega}$$

are fulfilled with M as set of all meshes and $\bar{\lambda}$ as *mean way length* or *mean mesh length*.

Theorem 2.2.-2 (Euler's theorem): Let Σ_2 be a connected oriented neighbor hood structure $[P,N,C]$ with ε points, $\varkappa$ edges and ω meshes. Then

$$\varepsilon - \varkappa + \omega = 2(1-g) = \chi \le 2$$

is fulfilled with g as *genus* of the structure and χ as *Euler characteristic* of the structure.

Theorem 2.2.-3 (substructure theorem): If Σ' is a substructure of a finite connected structure Σ, then $\chi(\Sigma')\geq\chi(\Sigma)$. If the structure Σ consists of k connected components then $\chi(\Sigma)=\varepsilon-\varkappa+\omega\leq 2k$.

Theorem 2.2.-4 (mesh number theorem): A given neighborhood structure Σ belongs independently of its orientation either in the structure class characterized by an even mesh number ω or in the structure class characterized by an odd mesh number ω.

Theorem 2.2.-6 (border mesh theorem): Let Σ_M be a proper substructure of a finite structure $\Sigma=[P,N,C]$ with $M\subset P$, and let the structure Σ' be constructed out of Σ by deleting of all boundary edges attached to a certain border mesh of Σ_M. Then it is always $\chi(\Sigma')>\chi(\Sigma)$.

Theorem 2.2.-7: Let Σ' be constructed out of a finite connected structure Σ by deleting of all boundary edges attached to a certain border mesh of Σ. Assuming that $\chi(\Sigma)=2$, i.e. Σ is a so-called *planar structure*, $\chi(\Sigma')>2$ follows, and Σ' consists of at least two components.

Theorem 2.2.-9 (theorem of minimal color number): The necessary number $c_{\min}$ of colors for the regular coloring of a connected structures $\Sigma=[P,N,C]$ with an Euler characteristic $\chi=\chi(\Sigma)$ is given by

$$c_{\min}(\chi) = ent\left[\frac{7+\sqrt{49-24\chi}}{2}\right] .$$

Theorem 2.2.-11 (theorem of maximal color number): Each structure Σ can be regularly colored with at most

$$c_{\max}(\chi(\Sigma)) = ent\left[\frac{7+\sqrt{49-24\chi}}{2}\right]$$

different colors.

Formula 2.3.-1 (toroidal formula):

$$\frac{2}{\nu}+\frac{2}{\lambda}=1$$

Formula 2.3.-2 (Pick's formula):

$$\varepsilon=\frac{\lambda}{\nu}\mu+\frac{l}{2}+1$$

Theorem 2.3.-1 (total curvature theorem): In finite toroidal neighborhood structures $\Gamma(\nu,\lambda)$, the relationship

$$t = \frac{k}{\nu} - \frac{l}{\lambda} \in \{-1, +1\}$$

is fulfilled for each border mesh with $t=+1$ for object conturs (outer border meshes), and with $t=-1$ for inner border meshes (conturs of holes).

Formula 2.4.-1 (object matching formula):

$$\begin{aligned} a_l b_{lk} - c_{lk} &= a_k b_{kl} && \text{for} \quad k > l \;, \\ a_l b_{lk} &= a_k b_{kl} && \text{for} \quad k = l \;, \\ a_l b_{lk} + c_{kl} &= a_k b_{kl} && \text{for} \quad k < l \;. \end{aligned}$$

Formula 2.4.-2 (identity formula):

$$-\sum_{k=m+1}^{n} (-1)^k \binom{n}{k} = \sum_{k=0}^{m} (-1)^k \binom{n}{k} = (-1)^m \binom{n-1}{m}$$

Formula 2.4.-3 (boundary formula):

$$\psi^{(n)} = \frac{1}{2n} \sum_{l=1}^{n} (-1)^{l+1} c_{l-1,l}$$

Formula 2.4.-4 (object quantities):

$$q_m = \sum_{j=0}^{n-m} (-1)^j \binom{j+m}{j} a_{j+m}$$

Formula 2.4.-5 (geometric main formula):

$$q_m^{(n)} = \frac{1}{2(n-m)} \sum_{l=m+1}^{n} (-1)^{l-m+1} \binom{l-1}{m} c_{l-1,l}$$

Formula 2.4.-6 (two-dimensional form factor):

$$f = \frac{q_1^2}{4 q_2} = \frac{(\varkappa - 2\mu)^2}{4\mu} = \frac{I^2}{16\mu} \geq 1$$

Formula 2.4.-7 (three-dimensional form factors):

$$g = \frac{q_1^2}{3 q_2} = \frac{(\varkappa - 2\mu + 3\zeta)^2}{3(\mu - 3\zeta)} = \frac{(c_{12} - 2 c_{23})^2}{24 c_{23}} \geq 1 ,$$

$$h = \frac{q_1 q_2}{9 q_3} = \frac{(\varkappa - 2\mu + 3\zeta)(\mu - 3\zeta)}{9\zeta} = \frac{(c_{12} - 2 c_{23})\, c_{23}}{72\zeta} \geq 1 .$$

Theorem 2.4.-1 (Minkowski's quermass-integrals): The n-dimensional volume of the outer parallel set C_d of a convex set C with distance d between C and C_d is a linear combination of the quermasses $M_k(C)$. The quermasses are defined by the formula

$$V(C_d) = \sum_{k=0}^{n} \binom{n}{k} M_k(C)\, d^k .$$

Formula 3.2.-1 (intersection formula):

$$S_k(B, B') = N_k^{(n)}\, a_k(B)\, a_k(B') = 2^{n-1}\, k!\, (n-k)!\, a_k(B)\, a_k(B')$$

Formula 3.2.-2 (generalized intersection formula):

$$S_R(B, B') = \sum_{k=0}^{n} r_k\, S_k(B, B') = \sum_{k=0}^{n} r_k\, N_k^{(n)}\, a_k(B)\, a_k(B')$$

$$= \sum_{k=0}^{n} r_k\, 2^{n-1}\, k!\, (n-k)!\, a_k(B)\, a_k(B')$$

Formula 3.2.-3 ($\psi^{(2)}$ count formula):

$$\psi^{(2)} = \sum_{x_p=1} \left[\overline{x_1}\,\overline{x_2} - x_1 x_2 \overline{x_3} \right]$$

Formula 3.2.-4 ($\psi^{(3)}$ count formula):

$$\psi^{(3)} = \sum_{x_p=1} \left[\overline{x_1}\,\overline{x_2}\,\overline{x_3} + (1 - \overline{x_1}\,\overline{x_2}\,\overline{x_3})(F - G)\right] ,$$

$$F = (1 - x_7 x_4 x_5 x_6)(x_1 x_2 x_3 - x_1 x_2 \overline{x_4} - x_1 x_3 \overline{x_5} - x_2 x_3 \overline{x_6}) ,$$

$$G = x_7(1 - \overline{x_4}\,\overline{x_5}\,\overline{x_6})(x_1 x_2 \overline{x_3}\,\overline{x_4} + x_1 \overline{x_2} x_3 \overline{x_5} + \overline{x_1} x_2 x_3 \overline{x_6}) .$$

Formula 3.2.-5 ($\psi^{(2)}$ expectation value formula):

$$E(\psi^{(2)}) = \alpha\left(p - \frac{\nu}{2}p^2 + \frac{\nu}{\lambda}p^{\lambda}\right)$$

Formula 3.2.-6 ($\psi^{(n)}$ expectation value formula):

$$E(\psi^{(n)}) = \alpha \sum_{k=0}^{n} (-1)^k \binom{n}{k} p^{2k}$$

Theorem 4.1.-2 (Blichfeldt's theorem): Given a figure F in Euclidean space E^2 with area $A > N$ whereby N is an integer number. It is possible by translation that the translated figure F' covers at least $N+1$ lattice points.

Formula 4.1.-1 (digitalization effects):

$$E(N_S) = A(S) \quad , \quad \delta N_S = \sqrt{E(D)} = \sqrt{\frac{P(S)}{4\pi}} .$$

Formula 4.2.-1 (splitting formula):

$$\frac{\Delta x}{\Delta y} = [a_1, a_2, \ldots, a_n] = \cfrac{1}{a_1 + \cfrac{1}{a_2 + \cfrac{1}{\ddots \, \cfrac{1}{a_{n-1} + \cfrac{1}{a_n}}}}}$$

$$[a_1, a_2, \ldots, a_n] = [a_1, \ldots, a_{n-1} + 1] \oplus (a_n - 1) \cdot [a_1, \ldots, a_{n-1}]$$

Formula 4.3.-1 (maximal convex object formula):

$$m_{approx}(c) = 0.1507\, c^{3/2} + 0.1 + \frac{2}{c^{1/2}} ,$$

$$c_{approx}(m) = 3.53\, (m-1)^{2/3} .$$

Formula 4.4.-1 (shear transformation):

$$\begin{pmatrix} \cos\varphi & \sin\varphi \\ -\sin\varphi & \cos\varphi \end{pmatrix} = \begin{pmatrix} 1 & 0 \\ -\tan(\varphi/2) & 1 \end{pmatrix} \begin{pmatrix} 1 & \sin\varphi \\ 0 & 1 \end{pmatrix} \begin{pmatrix} 1 & 0 \\ -\tan(\varphi/2) & 1 \end{pmatrix}$$

Theorem 5.1.-1 (LSI operator theorem): An operator in function space $\boldsymbol{F}$ is a linear and shift-invariant operator (*LSI-operator*), if and only if it is a convolution operator.

Formula 5.2.-1 (pseudo scalar product):

$$\langle F*G, H\rangle = \sum_{p\in P} (F*G)_p H_p = \langle G, F^{\leftrightarrow} * H\rangle$$

Formula 5.2.-2 (discrete convolution theorem):

$$F*G = \sum_{p\in P} \varphi_p C^{(p)} * \sum_{q\in P} \gamma_q C^{(q)} = \sum_{p\in P} \varphi_p \gamma_p C^{(p)}$$

Theorem 5.2.-3 (orthogonal convolution base): An OCB exists in $\boldsymbol{F}$ if and only if the following requirements are fulfilled:

$$a^N = e ,$$

$$(a^k - e)^{-1} \quad \text{exists for} \quad k \in \{1, \ldots, N-1\} ,$$

$$N^{-1} \quad \text{exists} .$$

The algebraic quantity $a \in R$ is a primitive N-th root of the unit element e of R, and the expressions $a^k - 1$ are invertible in R.

Theorem 5.3.-1 (OCB function theorem): The base functions $C^{(k)} = (C_0^{(k)}, \ldots, C_{N-1}^{(k)})$ of an orthogonal convolution base are given by

$$C^{(k)} = \frac{1}{N} (e, \mathbf{a}^k, \mathbf{a}^{2k}, \ldots, \mathbf{a}^{(N-1)k})$$

so that $C^{(k)} * C^{(l)} = \delta_{k-l}\, C^{(k)}$ is fulfilled.

Formula 5.3.-1 (generalized Fourier transform):

$$F_n = N^{-1} \sum_{k=0}^{N-1} \varphi_k \mathbf{a}^{kn} \quad , \quad \varphi_k = \sum_{n=0}^{N-1} F_n \mathbf{a}^{-kn} .$$

Theorem 5.3.-2 (eigen values of OCB functions): For each discrete periodical function $F \epsilon \boldsymbol{F}$, the OCB functions $C^{(k)}$ are eigen functions of the convolution operator $\boldsymbol{C_F}$ with eigen values φ_k, i.e.

$$F * C^{(k)} = \varphi_k C^{(k)} .$$

Formula 5.3.-2 (Fourier transform):

$$F_n = N^{-1} \sum_{k=0}^{N-1} \varphi_k e^{\frac{2\pi i \cdot kn}{N}} \quad , \quad \varphi = \sum_{n=0}^{N-1} F_n e^{-\frac{2\pi i \cdot kn}{N}} .$$

Theorem 5.4.-1 (existence of inverse functions): The simple condition

$$\sum_{i=1}^{N-1} |H_i| < |H_0|$$

is sufficient that the inverse H^{-1} of function $H=(H_0,H_1,\ldots,H_{N-1})$ exists with regard to the operation *.

Formula 5.4.-1 (locally restricted convolution inverse):

$$\sum_{p \in Q \subset P} (F * F^{\leftarrow})_{p-q} X_p = F_q^{\leftrightarrow} \quad \text{for} \quad q \in Q$$

The following synopsis contains that symbols which are of more than local meaning.

$\boldsymbol{a}$	algebraic quantity of an orthogonal convolution basis
A	attachment relation: irreflexive, symmetrical
a_k	number of k-dimensional elements of an incidence structure
$a_k(B)$	number of k-dimensional elements of an object B
A_{ij}	matrix of adjacency
B	boundary relation: irreflexive, anti-symmetrical, transitive
b_{kl}	structure constant of homogeneous incidence structure
$B(M)$	set of all boundary points of M as subset of a metric space
$\boldsymbol{C}$	complementary operator with $\boldsymbol{C}M = \overline{M}$
$card(M)$	cardinality of set M, number of elements of set M
$c_{kl}(B)$	number of l-dimensional marginal elements attached to k-dimensional object elements of object B
$c(p)$	neighborhood cycle of a point p
C_F	circulant matrix corresponding to convolution with function F
$\boldsymbol{C_F}$	convolution operator corresponding to function F
$C^{(k)}$	function of an orthogonal convolution base (OCB)
$\boldsymbol{D}$	dilatation operator with $\boldsymbol{D}M = M \cup N(M)$
$d(p,q)$	metric distance between two points
$d \mid N$	d is divisor of N
e	unit element of an algebraic structure
$\boldsymbol{E}$	erosion operator with $\boldsymbol{E}M = K(M)$
$E(f)$	expectation value of random variable f
$E^{(k)}$	function of the canonical function base
$ent(x)$	entier-function (largest integer smaller than or equal to x)
$F^{\mid k}$	k-fold shifted function with $F_n^{\mid k} = F_{n-k}$
$F^{\leftrightarrow}$	reflected function with $F_n^{\leftrightarrow} = F_{-n}$
$\mathcal{F}$	function space
$\boldsymbol{F}$	Fourier operator
$gcd(i,j\ldots)$	greatest common divisor of integers $i,j,\ldots$
I	incidence relation: reflexive, symmetrical
$\boldsymbol{I}$	identity operator with $\boldsymbol{I}M = M$

$\boldsymbol{K}$	core building operator with $\boldsymbol{K}M=K(M)$
k	number of marginal edges of a two-dimensional object
$K(M)$	set of all core points of M as subset of a neighborhood structure
l	number of marginal meshes of a two-dimensional object
$\boldsymbol{L}$	linear operator
M^*	set of all subsets of set M
M^n	n-fold Cartesian product of set M
$M \times N$	Cartesian product of two sets M and N
$\overline{\boldsymbol{M}}$	complementary set
N	neighborhood relation: irreflexive, symmetrical
N	period of a one-dimensional discrete function
$\boldsymbol{N}$	neighborhood operator with $\boldsymbol{N}M=N(M)$
n	dimension of an incidence structure
$N(p)$	neighborhood of a point p
$N(M)$	set of all points neighbored to $M \subset P$ of a neighborhood structure
o	zero element of an algebraical structure
O	symbol for order of quantity or order of complexity
$\boldsymbol{O}$	opening operator with $\boldsymbol{O}M=K(M) \cup N(K(M))$
$p \sim q$	elements in equivalence relation or in connectivity relation
q_m	object quantity of magnification order m
$\boldsymbol{R}$	boundary building operator with $\boldsymbol{R}M=R(M)$
$\boldsymbol{R}$	reflection operator
$R(M)$	set of all boundary points of $M \subset P$ of a neighborhood structure
$\boldsymbol{S}$	closing operator with $\boldsymbol{S}M=K(M \cup N(M))$
$\boldsymbol{S}$	shift operator
$scm(i,j,\ldots)$	smallest common multiple of integers $i,j,\ldots$
$S_R(B,B')$	sum of function R over all intersections $B \cap B'$
$S_\alpha(B,B')$	sum of α-values over all intersections $B \cap B'$
$S_\delta(x)$	δ-surrounding of element x in a metric space
$\boldsymbol{S}^{\mid k}$	multiple shift operator
t	total curvature of a border mesh
$\boldsymbol{T}$	translation operator
T_p	translation

V	connectivity relation: reflexive, symmetrical, transitive
$\mathbf{Z}$	zero operator with $\mathbf{Z}M = \varnothing$
Z^n	n-dimensional orthogonal lattice (infinite or toroidally closed)
β	number of all boundary meshes of a two-dimensional object
$\Gamma(\nu,\lambda)$	two-dimensional finite homogeneous neighborhood structure
δ_k	Kronecker symbol
ε	number of all points of a graph or a neighborhood structure
ζ	number of all cells in a three-dimensional incidence structure
$\varkappa$	number of all edges of a graph or a neighborhood structure
$\lambda(m)$	length of a mesh m in a neighborhood structure
μ	number of all core meshes of an object in a neighborhood structure
$\nu(p)$	number of neighbors of a point p
Σ_d	d-dimensional incidence structure
$\varphi(k)$	Euler's divisor function
$\chi(\Sigma)$	Euler characteristic of an incidence structure Σ
$\psi(B)$	Euler number of an object B
ω	number of all directed ways (meshes) of a two-dimensional object
$\varnothing$	empty set
$\times$	set product operator
$*$	convolution operator
$\{\ldots\}$	sets
$(\ldots)$	sequences, ordered sets of elements of one kind
$[\ldots]$	structures, ordered sets of elements of different kinds
$<\ldots>$	cycles, cyclically ordered sets
$<F,G>$	(pseudo) scalar product of two functions

7 References

Ac75 N.I.Achieser, I.M.Glasmann: Theorie der linearen Operatoren im Hilbertraum. Akademie-Verlag, Berlin 1975

Ag75 R.C.Agarwal, C.S.Burrns: Number theoretic transforms to implement fast digital convolution. Proc. IEEE 63 (1975) 550-560

Ag85 G.J.Agin: Vision systems. In: S.Y.Nof (ed.): Handbook of Industrial Robotics. John Wiley & Sons, New York 1985, pp. 231-261

Ah78 N.Ahuja: Connectivity in lattices and mosaics. Techn.Rep. TR-637, Univ. of Maryland, Center for Automation Research, College Park, Feb. 1978

Ah78a N.Ahuja: Connectivity in lattices and mosaics - II. Techn.Rep. TR-671, Univ. of Maryland, Center for Automation Research, College Park, June 1978

Al56 P.S.Alexandroff: Einführung in die Mengenlehre und die Theorie der reellen Funktionen. Deutscher Verlag der Wissenschaften, Berlin 1956

An85 S.Ansaldi, L.De Floriani, B. Falcidieno: Geometric modelling of solid objects by using a face adjacency graph representation. Comput.Graph. 19 (1985) 131- 139

Ap76 K.Appel, W.Haken: Every planar map is four colorable. Bull. Amer. Math. Soc. 82 (1976), No. 5

Ar72 C.Arcelli, S.Levialdi: Parallel shrinking in three dimensions. CGIP 1 (1972) 21-30

Ar81 E.Artzy, G.Frieder, G.T.Herman: The theory, design, implementation and evaluation of a three-dimensional surface detection algorithm. CGIP 15 (1981) 1-24

Ar81a E.Artzy, G.T.Herman: Boundary detection in 3 dimensions with a medical application. Computer Graphics 15 (1981) 92-123

Ba67 R.M.Baer, D.W.Matula: Circles in a square world. Comp. Biomed. Res. 1 (1967) 18-40

Ba74 B.Baumgart: Geometric modelling for computer vision. Techn. Rep. MEMO AIM-249, Stanford Artif.Intell Lab., Stanford University, Oct. 1974

Be79 L.Berg: Differenzengleichungen zweiter Ordnung mit Anwendungen. Deutscher Verlag der Wissenschaften, Berlin 1979

Be82 A.Beutelspacher: Einführung in die endliche Geometrie.I. Wissenschaftsverlag Mannheim/Wien/Zürich 1982

Be85 U.Betke, P.McMullen: Lattice points in lattice polytopes. Monatsh. Math. 99 (1985) 253-265

Be85a U.Betke, M.Kneser: Zerlegungen und Bewertungen von Gitterpolytopen. J.reine u. angew. Mathem. 358 (1985) 202-208

Be88 C.A.Berenstein, D.Lavine: On the number of digital straight line segments. IEEE Trans. PAMI-10 (1988) 880-887

Bi84 H.Bieri, W.Nef: Algorithms for the Euler characteristics and related additive functionals of digital objects. CVGIP 28 (1984) 166-175

Bi87 H.Bieri: Computing the Euler characteristic and related additive functionals of digital objects from their bintree representation. CVGIP 40 (1987) 115-126

Bl14 H.F.Blichfeldt: A new principle in the geometry of numbers with some applications. Trans. Amer. Math. Soc. 15 (1914) 227-235

Bl21 H.F.Blichfeldt: Note on geometry of numbers. Bull. Amer. Math. Soc. 27 (1921) 152-153

Bl55 W.Blaschke: Vorlesungen über Integralgeometrie. Deutscher Verlag der Wissenschaften, Berlin 1955

Bo34 J.Bonnesen, V.Fenchel: Theorie der konvexen Körper. Springer Verlag, Berlin 1934

Bo71 W.G.Boltjanski, I.M.Jaglom: Konvexe Figuren und Körper. In: P.S.Alexandroff, A.I.Markuschewitsch, A.J.Chintschin (Hrsg.): Enzyklopädie der Elementarmathematik. Bd.5, Deutscher Verlag der Wissenschaften, Berlin 1971, S.173-257

Bo72 J.Bokowski, H.Hadwiger, J.M.Wills: Eine Ungleichung zwischen Volumen, Oberfläche und Gitterpunktanzahl konvexer Körper im n-dimensionalen euklidischen Raum. Math. Zeitschr. 127 (1972) 363-364

Bo79 B.Bollobas: Graph Theory - An Introductory Course. Springer Verlag Berlin/ Heidelberg 1979

Bo81 K.S.Booth, R.J.Lipton: Computing extremal and approximate distances in graphs having unit cost edges. Acta Informatica 15 (1981) 319-328

Bö87 J.Böhm, E.Hertel: Jenaer Beiträge zur diskreten Geometrie. Wiss.Zeitschr. FSU, Heft 1, Jena 1987

Bö91 J.Böhm, E.Quaisser: Schönheit und Harmonie geometrischer Formen. Akademie-Verlag, Berlin 1991

Br41 R.L.Brooks: On colouring the nodes of a network. Proc. Cambridge Phil.Soc. 37 (1941) 194-197

Br80 I.Braid, R.Hillyard, I.Stroud: Stepwise construction of polyhedra in geometric modelling. In: Mathematical Methods in Computer Graphics and Design. Academic Press, London 1980, pp. 123-141

Br91 A.M.Bruckstein: Self-similarity properties of digitized straight lines. Contemp. Mathem. 119 (1991) 1-20

Bu85 W.Bunse: Numerische lineare Algebra. Teubner-Verlag, Stuttgart 1985

Bu88 P.Bundschuh: Einführung in die Zahlentheorie. Springer-Verlag, Berlin/ Heidelberg 1988

Bu91 A.Bunde, S.Havlin: Percolation I. In: A.Bunde, S.Havlin (eds.): Fractals and Disordered Systems. Springer Verlag, Berlin/ Heidelberg 1991, pp. 51-95

Ca59 J.W.S.Cassels: An Introduction to the Geometry of Numbers. Springer-Verlag, Berlin/Heidelberg 1959

Ca80 E.Catmul, A.R.Smith: 3-D transformation of images in scanline order. ACM SIGGRAPH '80, 1980, pp. 279-285

Ca85 M.Campanino, L.Russo: An upperbound for the critical probability for the three-dimensional cubic lattice. Ann.Prob. 13 (1985) 315-323

Ch82 R.Chellappa, R.L.Kashyap: Texture synthesis using spatial interaction models. Proc. IEEE Comp.Soc.Conf. PRIP, Las Vegas 1982, pp. 226-230

Ch85 L.S.Chen, G.T.Herman, R.A.Reynolds, J.K.Udupa: Surface rendering in the cuberille environment. IEEE Comp.Graph.Appl. 5 (1985) 33-43

Ch87 R.T.Chin, H.K.Wan: A one-pass thinning algorithm and its parallel implementation. CVGIP 40 (1987) 30-40

Cl63 W.K.Clifford: The Postulates of the Science of Space. In: Gateway to the Great Books, Vol. 9, Encyclopaedia Britannica Inc. 1963

Co88 J.H.Conway, N.J.A.Sloane: Certain important lattices and their properties. In: J.H.Conway, N.J.A.Sloane (eds.): Sphere Packings, Lattices and Groups. Springer Verlag Berlin/Heidelberg 1988, pp. 94-135

Cr86 R.Creutzburg: Finite Signalfaltungen und finite Signaltransformationen in endlichen kommutativen Ringen mit Einselement. Dissertation 1984, Wilhelm-Pieck-Universität Rostock (ZKI-Informationen 2/1986, Zentralinst. Kybernetik u. Inform.Prozesse Berlin)

Da51 H.Davenport: On a principle of Lipschitz. J. London Math. Soc. 26 (1951) 179-183

Da67 G.B.Danzig: All shortest routes in a graph. Proc.Int.Sympos. Theory of Graphs, Rome 1966, New York 1967, pp.91-92

Da81 E.R.Davies, A.P.N.Plummer: Thinning algorithms - a critique and a new methodology. Pattern Recognition 14 (1981) 53-63

Da88 A.M.Day: planar convex hull algorithms in theory and practice. Comp. Graph. Forum 7 (1988) 177-193

Da90 A.M.Day: The implementation of an algorithm to find the convex hull of a set of three-dimensional points. ACM Trans.Graphics 9 (1990) 105-132

Di59 E.W.Dijkstra: A note on two problems in connection with graphs. Numer.Math. 1 (1959) 269-271

Do84 L.Dorst, R.P.W.Duin: Spirograph theory - a framework for calculation on digitized straight lines. IEEE Trans. PAMI-6 (1984) 632-639

Do86 L.Dorst: Discrete Straight Line Segments. Ph.D. Thesis, Technological University Delft, 1986

Do87 D.P.Dobkin, M.J.Laszlo: Primitives for the manipulation of three-dimensional subdivisions. Proc. 3rd Ann.Sympos. on Computational Geometry. Waterloo, June 1987, pp.86-99

Do89 L.Dorst: Discrete straight line segments - parameters, primitives, and properties. In: A.Hübler, W.Nagel, B.D.Ripley, G.Werner (eds.): Geobild'89. Mathem.Res. Vol. 51, Akademie-Verlag, Berlin 1989, pp. 11-18

Du67 R.O.Duda, P.E.Hart, J.H.Munson: Graphical data processing research study and experimental investigation, AD 650926, March 1967, pp.28-30 (quoted in [Ko86a])

Du78 E.Dubois, A.N.Venetsanopoulos: The discrete Fourier transform over finite rings with applications to fast convolution. IEEE Trans. C-27 (1978) 586-593

Du80 E.Dubois, A.N.Venetsanopoulos: The generalized discrete Fourier transform in rings of algebraic integers. IEEE Trans. ASSP-28 (1980) 169-175

Du84 D.E.Dudgeon, R.M.Merserau: Multidimensional Digital Signal Processing. Prentice Hall 1984

Ec87 U.Eckardt: Digital topology. II.Perfect points on the inner boundary. Hamburger Beiträge Angew.Math. A, Preprint 11, Nov. 1987

Ec88 U.Eckhardt, G.Maderlechner: Parallel reduction of digital sets. Siemens Forsch./Entwickl.Ber. 17 (1988) 184-189

Ed87 H.Edelsbrunner: Algorithms in Combinatorical Geometry. Springer Verlag Berlin/Heidelberg 1987

Ei52 S.Eilenberg, N.Steenrod: Foundations of Algebraic Topology. Princeton 1952

Eu36 L.Euler: Solutio problematis ad geometriam situs pertinentis. Comment.Acad. Petropolit. 8 (1736) 128-140

Es80 J.W.Essam: Percolation theory. Reports on Progress in Physics 43 (1980) 833-912

Fr61 H.Freeman: On the encoding of arbitrary geometry configurations. IRE Trans. EC-10 (1961) 260-268

Fr71 H.Freeman: Boundary encoding and processing. In: S.Lipkin, A.Rosenfeld (eds.): Picture Processing and Psychopictories. Academic Press, New York 1971, pp. 241-266

Ga77 M.Gaafar: Convexity verification, block-chords, and digital straight lines. CGIP 6 (1977) 361-370

Ge91 G.Gerig, C.Brechbühler, P.Droz, O.Kübler: 3-D Verdünnung zur symbolischen Beschreibung von verästelten, räumlichen Strukturen. In: B.Radig (Hrsg.): Mustererkennung 1991. Proc. 13. DAGM-Sympos., Springer-Verlag, Berlin/Heidelberg 1991, S. 304-311

Gi77 P.J.Giblin: Graphs, Surfaces and Homology - An Introduction to Algebraic Topology. Chapman and Hall, London 1977

Go85 J.E.Goodman, E.Lutwack, J.Malkevitch, R.Pollack (eds.): Discrete Order and Convexity. Ann. NY Acad. Sc., Vol. 440, New York 1985

Go89 D.Gordon, J.K.Udupa: Fast surface tracking in three-dimensional binary images. CVGIP 45 (1989) 196-214

Gr71 S.B.Gray: Local properties of binary images in two and three dimensions. IEEE Trans. C-20 (1971) 551-561

Gr85 B.Grünbaum, G.C.Shephard: A catalogue of isonemal fabrics. In: J.E. Goodman, E.Lutwack, J.Malkevitch, R.Pollack (eds.): Discrete Order and Convexity. Ann. NY Acad. Sc., Vol. 440, New York 1985, pp. 279-298

Gr85a R.L.Graham: Old and new Euclidean Ramsey theorems. In: J.E. Goodman, E.Lutwack, J.Malkevitch, R.Pollack (eds.): Discrete Order and Convexity. Ann. NY Acad. Sc., Vol. 440, New York 1985, pp. 20-30

Gr86 D.H.Greene, F.F.Yao: Finite-resolution computational geometry. Proc. IEEE Annual Sympos. on Foundations of Computer Science 1986, pp. 143-152

Gr87 P.M.Gruber, C.G.Lekkerkerker: Geometry of Numbers. North-Holland, Amsterdam 1987

Gr89 G.R.Grimmett: Percolation. Springer-Verlag, Berlin/Heidelberg 1989

Ha14 F.Hausdorff: Grundzüge der Mengenlehre. Leipzig 1914

Ha64 H.Hadwiger, H.Debrunner: Combinatorical Geometry in the Plane. Holt/Rinehardt/Whinston, New York 1964

Ha68 P.L.Hammer, S.Rudeanu: Boolean Methods in Operation Research and Related Areas. Springer Verlag Berlin/Heidelberg 1968

Ha69 F.Harary: Graph Theory. Reading/Mass. 1969

Ha84 F.Harary, R.A.Melter, I.Tomescu: Digital metrics - a graph theoretical approach. Pattern Recognition Letters 2 (1984) 159-163

Ha89 R.Halin: Graphentheorie. Akademie-Verlag, Berlin 1989

He83 G.T.Herman, D.Webster: A topological proof of a surface-tracking algorithm. CVGIP 23 (1983) 162-177

He90 P.J.Heawood: Map colour theorem. Quart.J.Math. 24 (1890) 332-338

Hü81 A.Hübler, R.Klette, K.Voss: Determination of the convex hull of a finite set of planar points within linear time. Elektron. Inform. Verarb. Kybern. EIK 17 (1981) 121-139

Hü87 A.Hübler, R.Klette: Theoretische Grundlagen der digitalen Bildverarbeitung. VII - Geometrie auf Nachbarschaftsstrukturen. Bild und Ton 40 (1987) 229-235

Hü87a A.Hübler: Ein abstrakter Zugang zur digitalen Geometrie - Möglichkeiten und Grenzen. In: E.Paulus (Hrsg.): Mustererkennung 1987 - Proc. 9th DAGM-Sympos. Braunschweig, Sept.1987, S.151-155

Hü89 A.Hübler: Diskrete Geometrie für die digitale Bildverarbeitung. Habilitationsschrift, Friedrich-Schiller-Universität Jena 1989

Ik81 K.Ikeuchi: Recognition of 3D-objects using the extended Gaussian image. In: Proc. IJCAI 1981, pp. 595-600

Ik83 K.Ikeuchi: Determining attitude of object from needle map using extended Gaussian image. AI Memo No.714, Cambridge, MIT, AI Lab. 1983

Ir89 M.C.Irwin: Geometry of continued fractions. Amer.Math. Monthly 96 (1989) 696-703

Ja83 L.Janos, A.Rosenfeld: Digital connectedness - an algebraic approach. Pattern Recognition Letters 1 (1983) 135-139

Ja85 L.P.Jaroslavskij: Einführung in die digitale Bildverarbeitung. Deutscher Verlag der Wissenschaften, Berlin 1985

Jä87 K.Jänich: Topologie. Springer Verlag, Berlin/Heidelberg 1987

Ji88 X.Y.Jiang, H.Bunke: Segmentierung von Nadeldiagrammen von Objekten mit gekrümmten Oberflächen. In: H.Bunke, O.Kübler, P.Stucki (Hrsg.): Mustererkennung 1988, 10.DAGM-Sympos. 1988, Springer-Verlag Berlin/Heidelberg 1988, S. 255-261

Ji90 X.Y.Jiang, H.Bunke: Recognizing 3-D objects in needle maps. Proc. 10th ICPR 1990, pp. 237-239

Ji91 X.Y.Jiang, H.Bunke: Ein konturbasierter Ansatz zur Berechnung von Momenten. In: B.Radig (Hrsg.): Mustererkennung 1991. Proc. 13.DAGM-Sympos., Springer-Verlag, Berlin/Heidelberg 1991, S. 143-150

Ka76 F.Karteszi: Introduction to Finite Geometries. Budapest 1976

Ka86 F.Kaden: Graphmetriken und Isometrieprobleme zugehöriger Distanzgraphen. ZKI-Informationen 1/86, Akademie der Wissenschaften, Berlin 1986

Ka88 T.Kasvand: Total 8-connectivity on the square raster. Proc. 9th ICPR, Rome 1988, pp. 354-356

Ka89 W.Kasprzak: A two-step modelling algorithm for tomographic scenes. In: H.Burkhardt, K.H.Höhne, B.Neumann (Hrsg.): Mustererkennung 1989, Springer-Verlag Berlin/Heidelberg 1989, S. 119-123

Ke63 M.G.Kendall, P.G.P.Moran: Geometrical Probability. Griffin, London 1963, pp. 104

Ke82 H.Kesten: Percolation Theory for Mathematicians. Birkhäuser-Verlag, Boston/ Basel/Stuttgart 1982

Ki80 C.E.Kim: A linear time convex hull algorithm for simple polygons. Techn. Rep. TR-956, Univ. of Maryland, Comp.Sc. Center, College Park, Aug. 1980

Ki81 C.E.Kim: Three-dimensional digital straight line segments. Techn. Rep. TR-1087, Univ. of Maryland, Comp.Sc.Center, College Park, Aug. 1981

Ki82 C.E.Kim, A.Rosenfeld: Digital straight lines and convexity of digital regions. IEEE Trans. PAMI-4 (1982) 149-153

Ki82a C.E.Kim: Three-dimensional digital planes. Techn. Rep. TR-1190, Univ. of Maryland, Comp.Sc.Center, College Park, July 1982

Kl33 F.Klein: Elementarmathematik vom höheren Standpunkt aus. Bd.2. Springer Verlag Berlin/Heidelberg 1933

Kl72 F.Klein: Das Erlanger Programm (Erlangen Okt. 1872). In: H.Wußing (Hrsg.): Ostwalds Klassiker der exakten Wissenschaften, Bd. 253, Leipzig 1972

Kl83 R.Klette: M-dimensional cellular spaces. Techn. Rep. CAR-TR-6, Univ. of Maryland, Center for Automation Research, College Park, Aug. 1983

Kl83a R.Klette: The m-dimensional grid point space. Techn. Rep. TR-1256, Univ. of Maryland, Comp.Sc.Center, College Park, March 1983

Kl85 R.Klette, K.Voss, P.Hufnagl: Theoretische Grundlagen der digitalen Bildverarbeitung. II. Nachbarschaftsstrukturen. Bild und Ton 38 (1985) 325-331

Kl85a B.Klotzeck, U.Lengtat, E.Letzel, K.Schröter: Kombinieren, Parkettieren, Färben. Aulis-Verlag Deubner, Köln 1985

Kl86 R.Klette, K.Voss, P.Hufnagl: Theoretische Grundlagen der digitalen Bildverarbeitung. III. Gebietsnachbarschaftsstrukturen. Bild und Ton 39 (1986) 45-50

Kl87 R.Klette, K.Voss: The three formulae of oriented graphs. Techn. Rep. CAR-TR-305, Center for Automation Research, University of Maryland, July 1987

Kl87a R.Klette, K.Voss: The three basic formulas of neighborhood structures. In: L.P.Yaroslavskij, A.Rosenfeld, W.Wilhelmi (eds.): Computer Analysis of Images and Patterns. Proc. 2nd Int.Conf. CAIP'87, Akademie-Verlag, Berlin 1987, pp. 149-161

Ko84 V.A.Kovalevsky: Discrete topology and contour definition. Pattern Recognition Letters 2 (1984) 281-288

Ko86 D.König: Theorie der endlichen und unendlichen Graphen. Teubner-Archiv Math. Bd.6, Teubner Verlagsgesellschaft, Leipzig 1986

Ko86a T.Y.Kong, A.Rosenfeld: Digital topology - introduction and survey. Techn.Rep. CAR-TR-229, Center for Automation Research, University of Maryland, Oct. 1986

Ko87 V.A.Kovalevski: The topology of cellular complexes as applied to image processing. In: L.P.Yaroslavskij, A.Rosenfeld, W.Wilhelmi (eds.): Computer Analysis of Images and Patterns. Proc. 2nd Int.Conf. CAIP'87, Akademie-Verlag, Berlin 1987, pp. 162-173

Ko89 T.Y.Kong, A.Rosenfeld: Digital topology - introduction and survey. CVGIP 48 (1989) 357-393

Ko89a V.A.Kovalevski: Finite topology as applied to image analysis. CVGIP 46 (1989) 141-161

Ko90 V.A.Kovalevski: Topology of cell complexes and its application to geometrical computations. Preprint of special issue "Topology in Computer Graphics and Image Processing" of journal "Topology and Its Applications", Proc. Sympos. Louisville 1990

Kr81 E.Krätzel: Zahlentheorie. Deutscher Verlag der Wissenschaften, Berlin 1981

Ku30 C.Kuratowski: Sur le probleme des courbes gauches en topology. Fund.Math. 15 (1930) 271-283

Ku76 F.Kuhnert: Pseudoinverse Matrizen und die Methode der Regularisierung. Teubner-Verlag, Leipzig 1976

Le52 S.Lefschetz: Algebraic Topology. New York 1952

Le80 K.Leichtweiß: Konvexe Mengen. Deutscher Verlag der Wissenschaften, Berlin 1980

Le85 C.Lee, A.Rosenfeld: Continuous representations of digital images. Techn. Rep. CAR-TR-158, Center for Automation Research, University of Maryland, Oct. 1985

Le85a C.Lee,A.Rosenfeld: Holes and genus of 3D digital images. Techn.Rep. CAR-TR-170, Center for Automation Research, University of Maryland, Dec. 1985

Le86 C.N.Lee, A.Rosenfeld: Connectivity issues in 2D and 3D images. Computer Vision, Pattern Recognition 1986, pp. 278-285

Le86a C.N.Lee, A.Rosenfeld: Computing the Euler number of a 3D image. Techn. Rep. CAR-TR-205, Center for Automation Research, University of Maryland, May 1986

Le87 C.N.Lee, A.Rosenfeld: Computing the Euler number of a 3D image. Proc. 1st Int. Conf. Computer Vision, London, June 1987, pp. 567-571

Li77 H.K.Liu: Two- and three-dimensional boundary detection. CGIP 6 (1977) 123-134

Li91 P.Lienhardt: Topological models for boundary representation - a comparison with n-dimensional generalized maps. Computer-Aided Design 23 (1991) 59-82

Lo80 S.Lobregt, P.W.Verbeek, F.C.A.Groen: Three-dimensional skeletonization - principle and algorithm. IEEE Trans. PAMI-2 (1980) 75-77

Lu89 M.Lucka, M.Vajtersic, R.Creutzburg, H.J.Grundmann: Parallel associative fast Fermat number transform implementation. Comp. Artif. Intell. 8 (1989) 267-280

Mä82 M.Mäntyla, R.Sulonen: GWB - a solid modeler with Euler operators. Comput. Graph. Appl. 2 (1982/7) 17-31

Mä83 M.Mäntyla: Computational topology - a study of topological manipulations and interrogations in computer graphics and geometric modeling. Acta Polytechnica Scandinavia 37 (1983)

Mc79 J.H.McClellan, C.M.Rader: Number Theory in Digital Signal Processing. Prentice Hall, London 1979

Mc77 P.McMullen: Valuations and Euler-type relations for polyhedral cones and polytopes. Proc. London Math. Soc. 35 (1977) 113-135

Me59 H.Meschkowski: Differenzengleichungen. Vandenhoek & Ruprecht, Göttingen 1959

Me89 R.Meester, T.Nowicki: Infinite clusters and critical values in two-dimensional circle percolation. Israel Journal of Mathematica 68 (1981) 63-81

Mi69 M.Minsky, S.Papert: Perceptrons. MIT Press, Cambridge 1969

Mi78 E.Minieka: Optimization Algorithms for Networks and Graphs. New York/Basel 1978

Mi96 H.Minkowski: Geometrie der Zahlen. Teubner-Verlag, Leipzig/Berlin 1896

Mo81 D.G.Morgenthaler: Three-dimensional simple points - serial erosion, parallel thinning, and skeletonization. Techn.Rep. TR-1005, Computer Science Center, University of Maryland, Febr. 1981

Mo81a D.G.Morgenthaler, A.Rosenfeld: Surfaces in three-dimensional digital images. Information and Control 51 (1981) 227-247

Mo85 W.O.Moser: Problems on extremal properties of a finite set of points. In: J.E.Goodman, E.Lutwack, J.Malkevitch, R.Pollack (eds.): Discrete Order and Convexity. Ann. NY Acad. Sc., Vol. 440, New York 1985, pp. 52-64

My71 J.P.Mylopoulos, T.Pavlidis: On the topological properties of quantized spaces. I. The notion of dimension. JACM 18 (1971) 239-246

Ni71 P.J.Nicholson: Algebraic theory of finite Fourier transforms. J.Comp. Syst. Sci. 5 (1971) 524-527

Ni86 J.Nievergelt, K.Hinrichs: Programmierung und Datenstrukturen. Springer Verlag, Berlin/Heidelberg 1986

Nö24 N.E.Nörlund: Vorlesungen über Differenzenrechnung. Springer-Verlag, Berlin 1924

No48 M.Nosarzweska: Evaluation de la difference entre l'aire d'une région plane convexe et le nombre des points aux coordonnées entières converts par elle. Coll.Math. 1 (1948) 305-311

Nu81 H.J.Nussbaumer: Fast Fourier Transforms and Convolution Algorithms. Springer Verlag, Berlin/Heidelberg 1981

Pa71 C.M.Park, A.Rosenfeld: Connectivity and genus in three dimensions. Techn.Rep. TR-156, Computer Science Center, University of Maryland, 1971

Pa77 T.Pavlidis: Structural Pattern Recognition. Springer Verlag, Berlin/Heidelberg 1977

Pa82 T.Pavlidis: Algorithms for Graphics and Image Processing. Springer Verlag, Berlin/Heidelberg 1982

Po04 H.Poincaré: Wissenschaft und Hypothese. Teubner-Verlag, Leipzig 1904

Po56 L.S.Pontrjagin: Grundzüge der kombinatorischen Topologie. Deutscher Verlag der Wissenschaften, Berlin 1956

Po71 I.H.Pollard: The fast Fourier transforms in a finite field. Math. Comput. 25 (1971) 365-374

Pr77 F.P.Preparata, S.J. Hong: Convex hull of a finite set of points in two and three dimensions. CACM 20 (1977) 87-93

Pr81 H.Primas: Chemistry, Quantum Mechanics, and Reductionism. Springer Verlag, Berlin/Heidelberg 1981

Pr85 F.P.Preparata, M.I.Shamos: Computational Geometry. Springer Verlag, Berlin/Heidelberg 1985

Qu79 B.v.Querenburg: Mengentheoretische Topologie. Springer Verlag, Berlin/ Heidelberg 1979

Ra72 C.H.Rader: Discrete convolutions via Mersenne transforms. IEEE Trans. C-21 (1972) 1269-1273

Re75 I.S.Reed, T.K.Truong: Complex integer convolutions over a direct sum of Galois fields. IEEE Trans. IT-21 (1975) 657-661

Ri59 G.Ringel: Färbungsprobleme auf Flächen und Graphen. Deutscher Verlag der Wissenschaften, Berlin 1959

Ri68 G.Ringel, J.W. Youngs: Solution of the Heawood map-coloring problem. Proc. Nat. Acad. Sci. USA 60 (1968) 438-445

Ri75 W.Rinow: Lehrbuch der Topologie. Deutscher Verlag der Wissenschaften, Berlin 1975

Ro66 A.Rosenfeld, J.L.Pfaltz: Sequential operations in digital picture processing. JACM 13 (1966) 471-494

Ro70 A.Rosenfeld: Connectivity in digital pictures. JACM 17 (1970) 146-160

Ro71 R.A.Rosenfeld, I.M.Jaglom: Nichteuklidische Geometrie. In: P.S.Alexandroff, A.I.Markuschewitsch, A.J.Chintschin (Hrsg.): Enzyklopädie der Elementarmathematik. Bd.5, Deutscher Verlag der Wissenschaften, Berlin 1971, S. 458-464

Ro74 A.Rosenfeld: Digital straight line segments. IEEE Trans. C-23 (1974) 1264-1269

Ro76 A.Rosenfeld, A.C.Kak: Digital Picture Processing. Academic Press, New York/London 1976

Ro76a J.Rothstein, C.Weiman: Parallel and sequential specification of a context sensitive language for straight lines on grids. CGIP 5 (1976) 106-124

Ro77 A.Rosenfeld: Digital topology. Techn. Rep. TR-542, University of Maryland, College Park, Computer Science Center, May 1977

Ro79 A.Rosenfeld: Digital topology. Amer.Math.Monthly 86 (1979) 621-630

Ro80 A.Rosenfeld, D.G.Morgenthaler: Some properties of digital curves and surfaces. Techn.Rep. TR-942, University of Maryland, College Park, Computer Science Center, Sept. 1980

Ro85 A.Rosenfeld: "Continuous" functions in digital pictures. Techn.Rep. CAR-TR-104, University of Maryland, College Park, Center for Automation Research, Jan. 1985

Ro87 A.Rosenfeld, R.A.Melter: Digital geometry. Techn.Rep. CAR-TR-323, University of Maryland, College Park, Center for Automation Research, Sept. 1987

Ro89 A.Rosenfeld, T.Y.Kong: Digital surfaces. Techn.Rep. CAR-TR-467, University of Maryland, College Park, Center for Automation Research, Oct. 1989

Ro89a C.Ronse: A bibliography on digital and computational convexity (1961-1988). IEEE Trans. PAMI-11 (1989) 181-190

Sa70 H.Sachs: Einführung in die Theorie der endlichen Graphen. I. Teubner Verlagsgesellschaft, Leipzig 1970

Sa72 H.Sachs: Einführung in die Theorie der endlichen Graphen. II. Teubner Verlagsgesellschaft, Leipzig 1972

Sa76 L.A.Santalo: Integral Geometry and Geometrical Probability. Addison-Wesley Publ.Comp., London 1976

Sa91 R.Safaee-Rad, V.C.Smith, B.Benhabib, I.Tchoukanov: Application of moment and Fourier descriptors to the accurate estimation of elliptical-shape parameters. Proc.Int.Conf. ASSP, Toronto, May 1991, Vol.4, pp. 2465-2468

Sc84 V.Schlegel: Nova Acta Kaiserl. Leopold. Carol. Akad., Bd. 44, Nr.4, 1884

Sc91 H.Scheid: Zahlentheorie. Wissenschaftsverlag Mannheim/Wien/ Zürich 1991

Sc91a M.Schubert: Texturanalyse mit dem autoregressiven Modell. In: B.Radig (Hrsg.): Mustererkennung 1991. Proc. 13.DAGM-Sympos. Springer Verlag Berlin/Heidelberg 1991, S. 231-238

Se68 J.Sedlacek: Einführung in die Graphentheorie. Teubner Verlagsgesellschaft, Leipzig 1968

Se78 P.D.Seymor, D.J.A.Welsh: Percolation probabilities on the square lattice. Ann. Discrete Math. 3 (1978) 227-245

Se84 J.Serra: Image Analysis and Mathematical Morphology. Academic Press, New York/London 1984

Sh90 M.Shizawa: Discrete invertible affine transformations. Proc. 10th ICPR, June 1990, Atlantic City, Vol. II, pp. 134-139

Si75 H.Simon, K.Kunze, K.Voss, W.R.Herrmann (Hrsg.): Automatische Bildverarbeitung in Medizin und Biologie. Steinkopff-Verlag, Dresden 1975

Sk70 J.Sklansky: Recognition of convex blobs. Pattern Recognition 2 (1970) 3-10

Sm61 W.I.Smirnow: Lehrgang der höheren Mathematik. Bd.4. Deutscher Verlag der Wissenschaften, Berlin 1961

Sp82 M.R.Spiegel: Endliche Differenzen und Differenzengleichungen. McGraw-Hill Book Comp. 1982

St79 D.Stauffer: Scaling theory of percolation clusters. Phys.Rep. 54 (1979) 1-74

St85 D.Stauffer: Introduction to Percolation Theory. Taylor and Francis, London 1985

St87 F.Stopp: Operatorenrechnung. Teubner-Verlagsgesellschaft, Leipzig 1987

St90 N.J.C.Strachan, P.Nesvadba, A.R.Allen: A method for working out the moments of a polygon using an integration technique. Pattern Recognition Letters 11 (1990) 351-354

Ta86 A.Tanaka, M.Kameyama, S.Kasama, O.Watanabe: A rotation method for raster image using skew transformation. Proc. IEEE Conf. on Computer Vision and Pattern Recognition, Miami 1986, pp. 272-277

Th82 R.Thiele: Leonhard Euler. Teubner Verlagsgesellschaft, Leipzig 1982

Th83 J.L.Thomas, G.N. Larsen, J.M.Keller: Number theoretic transforms with independent length and moduli. IEEE Trans. ASSP-31 (1983) 215-217

To73 G.Tourlakis, J.Mylopoulos: Some results in computational topology. JACM 20 (1973) 439-455

To82 J.Toriwaki, .Yokoi, T.Yonekura, T.Fukumura: Topological properties and topology-preserving transformation of a three-dimensional binary picture. Proc. IAPR Conf., Munich 1982, pp.414-419

Ts81 Y.F.Tsao, K.S.Fu: A parallel thinning algorithm for 3D pictures. CGIP 17 (1981) 315-331

Ud82 J.K.Udupa, S.N.Srihari, G.T.Herman: Boundary detection in multidimensions. IEEE Trans. PAMI-4 (1982) 41-50

Ud87 J.K.Udupa: A unified theory of objects and their boundaries in multi-dimensional digital images. Proc. Conf. CAR'87, pp. 779-784

Vo75 K.Voss: Digitalisierungseffekte in der automatischen Bildverarbeitung. Elektron. Inform.Verarb.Kybern. EIK 11 (1975) 469-477

Vo82 K.Voss, R.Klette: On the maximal number of edges of convex digital polygons included into a square. Comp. Artif. Intell. 1 (1982) 549-558

Vo85 K.Voss, P.Hufnagl, R.Klette: Theoretische Grundlagen der digitalen Bildverarbeitung. I. Einleitung. Bild und Ton 38 (1985) 299-303

Vo86 K.Voss, R.Klette: Theoretische Grundlagen der digitalen Bildverarbeitung. IV. Orientierte Nachbarschaftsstrukturen. Bild und Ton 39 (1986) 213-218

Vo86a K.Voss: Theoretische Grundlagen der digitalen Bildverarbeitung. V. Planare Strukturen und homogene Netze. Bild und Ton 39 (1986) 303-307

Vo87 K.Voss: Nachbarschaftsstrukturen als Bildträger. In: E.Paulus (Hrsg.): Mustererkennung 1987 - Proc. 9th DAGM-Sympos. Braunschweig, Sept. 1987, S.141-150

Vo87a K.Voss: Discrete integral geometry. In: L.P.Yaroslavskij, A.Rosenfeld, W.Wilhelmi (eds.): Computer Analysis of Images and Patterns. Akademie-Verlag, Berlin 1987, pp. 209-212

Vo87b K.Voss: Digitization effects in image analysis. Acta Stereologica 6/1 (1987) 145-147

Vo88 K.Voss: Theoretische Grundlagen der digitalen Bildverarbeitung. Akademie-Verlag, Berlin 1988

Vo88a K.Voss, M.Laudien: Faltungsinversion mittels Ausgleichsrechnung. In: Digitale Bildverarbeitung. Studientexte Techn. Univ. Dresden 1988, Bd. 104/88, S. 84-91

Vo89 K.Voss: Random point sets in lattices. In: A.Hübler, W.Nagel, B.D.Ripley, G.Werner (eds.): Geobild'89, Mathem.Research Vol. 51, Akademie-Verlag, Berlin 1989, pp. 25-28

Vo89a K.Voss, C.Hermann: Algebraic theory of discrete signals. In: K.Voss, D.Chetverikov, G.Sommer (eds.): Computer Analysis of Images and Patterns. Proc. 2nd CAIP, Leipzig 1989, Akademie-Verlag, Berlin 1989, pp. 77-82

Vo90 K.Voss: Detektion von Objektoberflächen in 3D-Bildern. Bild und Ton 43 (1990) 358-363

Vo91 K.Voss, H.Süße: Praktische Bildverarbeitung. Hanser-Verlag, München 1991

Vo91a K.Voss: Images, objects, and surfaces in Z^n. IJPRAI 5 (1991) 797-808

Vo91b K.Voss: Coding of digital straight lines by continued fractions. Comput. Artif. Intell. 10 (1991) 75-80

Vo91c K.Voss: An approach to algebraic theory of discrete signals. Comp. Artif. Intell. 10 (1991) 573-579

Vo92 K.Voss, I.Rothe: Characterization of objects by invariants. In: W.Förstner, S.Ruwiedel (eds.): Robust Computer Vision. Proc. Conf. on Quality of Vision Algorithms, Bonn, March 1992. Wichmann-Verlag, Karlsruhe 1992, pp.192-203

We76 C.F.R.Weiman: Highly parallel digitized geometric transformations without matrix multiplication. Proc. IEEE Int.Conf. on Parallel Processing 1976, pp. 1-10

Wi29 J.R.Wilton: The lattice points of a circle - an historical account of the problem. Mess.of Math. 58 (1929) 67-80

Wi90 J.M.Wills: Kugellagerungen und Konvexgeometrie. Jahresber. Deutsch. Math. Verein. 92 (1990) 21-46

Wi91 W.Wilhelmi: Image interpretation by algebraic topology. In: R.Klette (ed.): Computer Analysis of Images and Patterns. Akademie-Verlag, Berlin 1991, pp. 172-183

Wo80 R.J.Woodham: Photometric method for determining surface orientation from multiple images. Optical Engineering, Febr. 1980, pp. 139-144

Wu89 S.T.Wu: A new combinatorical model for boundary representation. Comput. & Graphics 13 (1989) 477-486

Za87 M.F.Zakaria, L.J. Vroomen, P.J.A. Zsomber-Murray, J.M.H. VanKessel: Fast algorithm for the computation of moment invariants. Pattern Recognition 20 (1987) 639-643

Za89 P.Zamperoni: Methoden der digitalen Bildsignalverarbeitung. Vieweg-Verlag, Braunschweig 1989

Эф82 A.L.Efros: Physics and Geometry of Unordering (Russian). Science Publ. Moscow 1982 / А.Л.Эфрос: Физика и Геометрия Беспорядка. Изд. Наука, Москва 1982

Яш65 P.P.Jaschinskas, R.Schomskis: Digitalization problems for two-dimensional images (Russian). In: Automation Control and Computer Technique. Mintis Publ., Vilnia 1965, pp. 5-10 / П.П. Яшинскас, Р.Шомскис: К вопросу о дискретизировании двухмерных изображенией. В: Автоматика и вычислительная техника. Изд. Минтис, Вильнюс 1965, с. 5-10

Translations of German quotations

T1) "I was once presented with a problem concerning an island in Königsberg surrounded by a river which is crossed by seven bridges. I was asked whether someone could cross each bridge on a continuous path so that each bridge would only be crossed once. At the same time I was also told that until now no-one had claimed this possible nor had anyone proved that it was impossible to do." [Euler, Ko86]

T2) "Graph theory can be interpreted from two different viewpoints. On the one hand, the theory is, as the theory of one-dimensional complexes, the first part of the general topology. On the other hand, the theory can be seen, if one abstracts from its continuous-geometric content, as a branch of combinatorics and abstract set theory. In this book, the second interpretation will be shown to advantage mainly because we ascribe absolutely no geometrical meaning to the elements of graphs, i.e. the points and edges: the points (nodal points) are all distinguishable elements, and an edge is nothing but a summation of its two end points." [König, Ko86]

T3) "A graph consists of a set X of elements of the first kind, the so-called nodal points, and a set U of elements of the second kind, as well as a function defined for U (the incidence function) which unambiguously assigns to every element $u \in U$ an ordered or unordered pair of (not necessarily different) nodal points $x,y \in X$." [Sachs, Sa70]

T4) "Let M be any finite or infinite set. To every (not necessarily ordered) pair (A,B) which can be constructed from elements of M, we assign a finite or infinite cardinality $\boldsymbol{m}_{AB}=\boldsymbol{m}_{BA}$ which can also be zero. If this assignment is so that for every A at least one $\boldsymbol{m}_{AB}$ is not zero, then a graph exists which possesses the elements of M as nodal points and in which any two nodal points A and B are joined by $\boldsymbol{m}_{AB}$ edges." [König, Ko86]

T5) "The following demands are generally made of a skeleton:

a) The skeleton must consist of lines the width of a single image point.
b) The topological connections of the skeleton must reflect those of the original image, i.e. the number of components connected in respect of the given metric and the Euler number must concur.
c) The skeleton-lines must pass approximately through the middle of the objects.
d) In the case of thick or eroded lines, as few as possible short branchings of the skeleton should occur.
e) The result of skeletonization may not further change after a certain number of iterations of the skeletonization algorithm." [Zamperoni, Za89]

T6) "The discrete nature of digital images makes it necessary to develop suitable definition systems and methods because a direct, immediate use of classical geometric theories is not possible and not adapted to the necessary requirements...The clear conception, going back to F.Klein [Kl72], of understanding geometry as the invariant theory of a given base set with regard to a transformation group is here not the determining element... We concur with the view of Böhm and Hertel [Bö87] which understands discrete geometry in sense of Klein's conception to be the invariant theory of a base set with a given system of figures in respect of a transformation." [Hübler, Hü89]

T7) Interesting is the conclusion that apart from 'linear' convex sets such as straight lines, rays and segments there are also 'higher-dimensional' convex point sets which are not connected...therefore one should look for ways to define more precisely the conception of convexity." [Hübler, Hü89]

8 Index

T

U

V

Z

Algorithms and Combinatorics

Editors: R. L. Graham, B. Korte, L. Lovász

Combinatorial mathematics has substantially influenced recent trends and developments in the theory of algorithms and its applications. Conversely, research on algorithms and their complexity has established new perspectives in discrete mathematics. This new series is devoted to the mathematics of these rapidly growing fields with special emphasis on their mutual interactions.

The series will cover areas in pure and applied mathematics as well as computer science, including: combinatorial and discrete optimization, polyhedral combinatorics, graph theory and its algorithmic aspects, network flows, matroids and their applications, algorithms in number theory, group theory etc., coding theory, algorithmic complexity of combinatorial problems, and combinatorial methods in computer science and related areas.

The main body of this series will be monographs ranging in level from first-year graduate up to advanced state-of-the-art research. The books will be conventionally type-set and bound in hard covers. In new and rapidly growing areas, collections of carefully edited monographic articles are also appropriate for this series. A subseries with the subtitle "Study and Research Texts" will be published in softcover and camera ready form. This will be mainly on outlet for seminar and lecture notes, drafts of textbooks with essential novelty in their presentation, and preliminary drafts of monographs. Refereed proceedings of meetings devoted to special topics may also be considered for this subseries. The main goals of this subseries are very rapid publication and the wide dissemination of new ideas.

Prospective readers of the series ALGORITHMS AND COMBINATORICS include scientists and graduate students working in discrete mathematics, operations research and computer science.

Volume 1: **K. H. Borgwardt**

The Simplex Method

A Probabilistic Analysis

1987. XI, 268 pp. 42 figs.
in 115 sep. illustrations.
ISBN 3-540-17096-0